TIME
OUT

A SECOND LOOK AT NATURE

CAL MBANO

ISBN   978-1-953821-34-8 Ebook
ISBN   978-1-953821-33-1 Paperback

The EC Publishing LLC books may be ordered
through booksellers or by contacting:

EC Publishing LLC
116 South Magnolia Ave.
Suite 3, Unit F
Ocala, FL 34471, USA
Direct Line: +1 (352) 644-6538
Fax: +1 (800) 483-1813
http://www.ecpublishingllc.com/

Ordering Information:
Quantity sales. Special discounts are available on quantity purchases by corporations, associations, and others. For details, contact the publisher at the address above.

Printed in the United States of America

ABOUT THE AUTHOR

Fig. 1: The Author

Callistus Mbano, Cal, is a keen lover of nature, mathematics, and science. His unique perspective of the world comes through a lens that looks deeply into the environment and delves into the everyday situations of life. Cal's distinct world view is broadened by his having traversed both time and space from the developing worlds to highly technological ones. He grew up in an indigenous Igbo village in the hinterland of Nigeria, survived the Biafran war, later became a civil engineer and teacher.

Cal earned his BS. Degree in Civil Engineering at the University of Michigan, Ann Arbor, MI; Master's Degree in Civil Engineering, Wayne State University, Detroit, MI, and Teaching Certification at the University of Phoenix, AZ. He is well-traveled and loves sharing his unique experiences with the people that he meets. He always has time for enjoying the simplicities of life with others, especially his four grandchildren. Take 'time out for a second look at nature' with Cal! You're sure to gain higher consciousness, curiosity, and appreciation for what nature teaches us about life.

Please visit or contact Cal on Facebook -*http://www.facebook.com/CalMbano*

Email: calm@thebabyboomerhub.com

DEDICATION

To my wife Cheryl Mbano,
whose contribution and encouragement
saw this work through.

KEYNOTE

Time Out: for a Second Look at Nature by Cal Mbano
explores the meaning of life,
the causes of unhappiness and how to restore joy to our world and maintain
wholeness from a naturalist perspective.

KEY WORDS

Creator, God, earth, heaven, religion, culture, nature, nurture, man, woman, Nigeria, America, British, Europeans, Ns Factor, evil, kindness, unhappiness, forest, solar system, recycling, animal, plants, food, outlier, terrorism, wars, Mahatma Gandhi, Martin Luther King, Nelson Mandela, IRA, PLO, Israel, Jews, Al-Qaeda, ISIS.

TABLE OF CONTENTS

INTRODUCTION

This book examines the state of the earth and why it's becoming increasingly intolerable —the evils and unhappiness are now the status quo for far too many people. Is it due to nature, nurture, or the combination of the two? Considering the advancements in science, technology - human knowledge - one would think that the earth is a much happier place. But did the Maker know this would be the case? Did He intend it to be so? **Time Out** considers these questions and proposes answers.

In our busy everyday living, we lack the consciousness of who we are, how our planet earth came to be our home, who put us on it, and why and our duties here on earth. This constant bedeviling amnesia induces neglect of our responsibilities: to take care of ourselves, one another, and the world. This state impacts us all since the consequences of unhappiness and the different evils we experience here have overwhelmed us. Some people have gone as far as denying the existence of the Creator as a result. Despite the mounting evidence to the contrary, those who persist on this route, including all the phenomena and wonders we experience daily throughout our lives, have themselves to blame.

I am not a psychologist, astrophysicist, or sociologist. My claim to science is only a master's degree in Civil Engineering. But I am an ardent lover of nature, the physical world, and a voracious reader. Time Out is about my observations of and interactions with the environment through which I became an unshakable believer in the Creator.

There is nothing spiritual about this work. There has been a lot of literature about the spirit, God. These writings can be found in the 'holy' books. I have a hard time understanding the metaphysical world because man, a creature bounded by time and space, can't think or reason in the realm of spirits. This statement in no way tries to contradict or negate the fact that the Almighty God can do as He pleases, including endowing some humans and animals with unique abilities to do things unnatural for their kinds.

Yes, that is our Creator, a spirit, all-powerful, with all knowledge and alone by Himself. Do these not explain why earthly and mortal humans cannot understand who the Creator is, leading them to rely on hear-say for knowledge and a leap of faith to believe in Him? This confusion resulted from early childhood instructions intending to provide insight and understanding of the Creator through the lens of ancient stories that have no contemporary significance. Hence, humans are in never-ending disobedience to His laws and incurring consequences as a result. How differently it could have been if humans were taught to study Nature - all created things in their environment - to observe God's writings to learn who the Creator is.

Since this writing is about frequent everyday observations available to an average person, I have taken care to keep it simple, in standard English Language with little scientific jargon. A friend wondered why a book of this theme does not contain citations, quotes, and examples from other literature on a similar subject. This book does not need all those because the information contained herein is self-evident. Distinct natural occurrences, though overlooked, need no supporting evidence or scientific proof. For example, a statement like: "Tomorrow will be another day" is self-evident.

The Designer and the owner of our earth and all the universe, all things visible and invisible, we call God in the English language; Chiukwu in Igbo (Ibo) language; Shen or Shangdi in the Chinese language; Allah in the Arabic language, etc. God can't go by one Universal name but by as many words as there are languages. It is confusing to hear some people argue that God has only one 'proper' and 'befitting' name. No! All these names refer to the Creator to those with open minds. Because of the lack of a universal term for the Creator, we use the following pronouns: He, Him, and His.

Nothing in His designs or creations shows that the Creator is either male or female. But considering all He created, the care and love showered on the earthly creations would make more sense to describe the Creator as female. Besides, through females, specifically mothers, their care and love, all living things we see continue to exist here on earth. Also, there have been great Queens who ruled their kingdoms, which means that the ruler-ship ability does not only belong to males. Also, consider the expressions "Mother Nature or Mother Earth" – all-female designations. So, there is no reason the Creator cannot be a female.

Then again, males are associated with power and strength. In the circumstances requiring power and strength, generally, the males of all the earthly creatures know it is their duty to deal with them. Think about the first-time humans encountered the oceans, seas, and even the rivers, stretching for miles with no end in sight. Those incredible creations struct them with fear and deep respect for the Creator. Think of the first Time those ancient peoples experienced thunder and lightning and even the Sun. They must have been struck similarly because of the strength and power of these creations. It is clear and understandable why they went as far as worshipping some of them.

In all these instances, what they saw was a show of strength and power, which are the dominant characteristics of males, males of all types of creatures. In ancient times, generally, males were heads of all institutions, starting with the families, the first institution humans knew. All these could explain why 'He' instead of 'she' is used to describe the Creator.

The phrase "Final Day" or "The Day of Judgement" is a common expression that needs no explanation. People use the two names interchangeably. It is assumed in some quarters that there will be a day, date, and time unknown to any human being (not the angels or any of the prophets). At the time, the Creator will call all human beings, dead or living, to appear before His Majesty and Glory to account for individual stewardship here on earth. At the end of this exercise, reward (heaven) or punishment (hell) will be meted accordingly. 'That Day' also suggests that there have been and will be periods of waiting for allocation following death. Those who have already died and those who will die before that 'Day' will have to stay somewhere, not in heaven (a place of enjoyment) or Hell (a place of punishment). Some suggest that the waiting will be in a halfway-house type of accommodation called 'Purgatory.'

There is nothing in the creation that alludes to the existence of that 'Day.' That expression is so ridiculous it should not merit further discussion. Why? The scenario suggests a lack of knowledge and hence, unpreparedness on the part of the Creator. How could this be? The Creator is All-Knowing!

That would be out of character for the Creator of all things, visible and invisible, the Omniscience. From examining all things around us, it is easy to understand that the 'Day' is referring to a time of transgression or the Time the Creator›s rule is disregarded. Therefore, there is no 'one' day but as many days as the transgressions. Also, reward or punishment is always proportional to the act and starts as soon as the deed was done.

A scenario will help to clarify this point: It might be difficult to believe that many of those seen as wealthy - having nothing to worry about, having all that money can buy – suffer incredibly. If these people open their private lives or explain how they amassed the wealth, many people will prefer poverty; leading me to conclude: "money or wealth can buy the best bed available on this earth' but can they also "buy a good night's sleep in the bed"? Think of people who have such a bed, but their eyes stay wide open throughout the night! Why? Guilt, the punishment for what they have done, has already started. The penalty is the 'Day.'

Also, consider time! Though intangible, time is as important as the Sun's energy, air, water, food, etc., in the lives of all living things. A human baby takes approximately nine months before it is ready to be born. Various times are needed with other animals. Eggs from different types of animals need time for readiness to hatch. Even the seeds of plants need time to become seedlings, produce fruits or become ready for eating. No child at birth can sit up, crawl, talk, reason, etc. These stages of life need time to become what they are created to be. Humans figured out that all activities here on earth require time to get their food and eat it. There is an element of time in all that living things do. It takes time to prepare food and to eat it.

At the beginning of time, humans worshiped some of God's creations like the Sun, the Rainbow, etc., because they did not know that those things were created by the same Creator that created them and did not possess any power over them. It was time that made the difference between their knowledge at their first encounters and the subsequent understanding of those creations.

Think of all the evils committed by humans because of lack of knowledge: At a time in some communities, twins were rejected as an evil omen; transgender and lesbian people were even killed because humans did not know better. But with time, they acquired the knowledge to understand that those conditions were as created by God.

It is inevitable then that human knowledge will continue to advance into sectors previously unimaginable. Inventions and discoveries will continue forever. Diseases and their causes will be better understood. The full potential of the medicinal values of plants and vegetables will be known and exploited, making it unnecessary to use chemicals to cure or manage diseases or enhance the quality of life, the soil, water, and environment. Then there will be no need to try to redesign, recheck the Creator's work. Humans will eventually understand the power, knowledge, and wisdom of the Creator. Time will make all these possible.

It makes sense to believe that everything will be in accord with the Creator; humans would have understood the meaning and full potential of Free Will and man's Ego. Then humans will realize the Creator's original intention - the earth will become the Paradise it was created to be.

The current state of the world continues, but 'soon' will run its course. Then understand why this world's evils – selfishness, cheating, murder, poverty, unhappiness, etc., defy solutions. Even advanced communities suffer the same human problems as less developed communities. Why? All created need time to become what they are designed to be. Time!

Time like Distance or Space is earthly. Time does not affect the Creator. Otherwise, the creation of the Universe would have been impossible. How long would it require to take the universe's physical measurement, count the number of stars, galaxies, etc., not to talk of creating each one? Also, while time is needed to get things done or things become realities, it has limiting effects – items don't reach fruition as fast as humans would want them to. Wouldn't humans desire to solve all human problems immediately, in no time? Yes! But it will never happen that way unless they figure out how to eliminate the Time factor.

A good question to ask would be whether there were other reasons why the Creator included time in the living things' environment. The orange seedling should become a matured fruiting tree if time is not involved. A baby will start walking as soon as it was born if time is absent. There is no reasonable answer to the question.

The Creator has His reasons unknown to humans. Will humans understand all the mysteries in creation with time? Only the Creator can answer that question. It makes sense to believe that secrets the Creator reserves to Himself, not revealed to His angels or the prophets.

Finally, I am asking for an open mind in reading this book. Why? You may find some statements rather unconventional, outside the norms. Any information repeated many times for an extended period becomes a 'fact' in the human mind, i.e., hard to shake off. It is true with many of our beliefs—human's way.

Right from the beginning of time, the Creator was explained to us by 'Holy People' or those who claimed to have been commissioned by Him, as a spirit being which He is. If the Creator appointed those people to teach us, why the confusion, the misconception of who the Creator is, leading to our disobeying His laws and the consequent never-ending suffering here on earth? What we should understand here is that humans are earthly creatures, not spirits. And baring divine intervention, humans cannot understand or reason in spirit.

Humans failed to understand they had no other choice than to study His creations, in their orbit, to understand who He is. The behavior of those who believe they could communicate with God in spirit proves this. These spirit believers, in extreme cases, are left with no other choice but to take a 'leap of faith.' It does not make sense that the way to understand or relate to the Creator is to take a 'leap' of faith! Such behavior is the reason for all the confusion and the consequences thereof.

Would it stand to reason that the Creator of our complicated Universe, the ever-knowing God, the Alpha, and Omega, would require or expect His earthly creatures to relate to or understand Him in spirit? That would be unusual and out of character. Indeed, He does as He pleases. There is nothing in our world that shows that was His intention. Instead, He left a lot of the extensions of Himself – His creations – that would render trying to understand Him in spirit, unnecessary. This book examines these extensions of Him in detail to try to explain some understanding of who the Creator is.

Chapter 1

REMINISCENCE

At a young age, I learned from my mother that I was 'created' by God. And that there was only one God that created everything (both the things that my eyes could see and the invisible ones). The response was a part of her answer to my childish question, "Who made me, you or my father." Then she inquired as to how I came about such a question.

When I was playing with my friends, we argued about who made children – the father or the mother? One boy said it was the father, but another said it was the mother. The oldest among us said it was both the mother and the father. I was confused because I was thinking of who provided the food; we always ate, which I assumed was my mother. My mind never strayed to thinking about the phrase, 'making a child.' I did not know children were 'made,' not to talk of who 'made' them.

My mother's answer created more questions and confusion in my tender mind. Has anybody seen God making anything before? Did any person see him making all these things? Did I not hear right that my father built our house? Or was it God that created it? My mother added, making me more troubled, that besides creating people, God also created everything. He was the most powerful and knew everything. Almost whispering, I asked her whether God was more powerful than my father, whom I thought, previously, was the greatest. When my mother answered yes to that question, the mother-son conversation ended abruptly. Someone was higher than my father! "Someone was greater than my father." How could that be possible!

Later in life, I was baptized, making me a Christian, and then I learned that there were three persons (beings) in that 'one' God, a concept known as the 'Holy Trinity.' I had never heard of such a description or seen three things that form one thing, and at the same time, the individual components existed as separate entities and equal in all respect.

When I thought about this three-in-one concept, I used to think about the Igbo soup my mother Prepared. Could that be like the Trinity? She used different ingredients to make this soup, like water, thickener, crayfish, meat, vegetables, oil, salt, and pepper. All these were boiled together, starting with water and the rest added at their proper times.

I saw the problem with this analogy. The obvious one was the number: There were 'three' persons in the Trinity but eight items in the soup. The figure did not bother me as much as the 'fact' that each entity stands on its own in the Trinity, and they are equal in all respect. But if you take a spoon of the Igbo soup, you cannot pick out the water, the thickener, the oil, the salt, and hardly the pepper. The original forms of the fish, meat, and vegetables had changed.

Chapter 2

THE QUESTION

The question, "How can three different entities, equal in all respect to each other, be one entity?" That question remains unanswered even today, notwithstanding the introduction of the new concept, Mystery. The situation remains the same because the new word "mystery" – a fact which the human mind could not comprehend did not help matters. However, my curiosity did not end there but was somewhat heightened.

As I observed all things around me and noted how complicated those things were in their natures, it became apparent that a being with supernatural powers and knowledge, God, created them, both the visible and invisible. The sheer number of things I saw and the incredible distances I came to know convinced me that it was God, not humans, that created everything. Yes, there must be a being with infinite powers and knowledge. Isn't it rather foolishness for anybody to believe that there is no God, that things created themselves, or just evolved by themselves?

Observing the complicated natures of created things in our orbit makes it clear God exists. But why all the wars, killings, poverty, different types of evils, and unhappiness, but only a few pockets of kindness? Wars and murders seemed to be a part of the human story.

The killing started when there were only a few people here on this earth, as legend has it: The story of a brother against his brother – Cain and Abel. Then came inter-tribal wars, as the population of the earth grew, and later, international battles. Finally, we participated in or hear stories of world wars. Now the killing has increased in frequency, magnitude, and sophistication. It is not restricted to one corner of the earth. No! It is seen everywhere. While you can observe poverty everywhere, it is more prevalent in less advanced economies.

Wars, killings, and poverty combined produced all sorts of evils and unhappiness. Modern wars are taking a new shape in the form of terrorism, cyber, and electronic warfare. Here the perpetrators of violence are invisible. There is no battle line. The world is a place of survival of the fittest. The rich get wealthier while the poor, poorer.

It is arguable to say that there is a fair amount of human kindness even in the face of the "Self-first" mentality. There have been some people who devoted their entire lives to the service of others. After amassing incredible wealth, many wealthy people spend the last years of their lives doing charity work. Yet, there is so much unhappiness, insecurity on this earth. Even the so-called rich are not immune to it, as evidenced by divorce, bitter family court battles, family breakups, suicides, etc. Why?

At this point, three more questions come to mind when one examines the state of this earth. First, is this what the Creator designed the world to be? Second, is the creation not complete? Or was there a mistake in the design? I am tempted to think that this state of the earth shows a disagreement among the 'three entities' in one God during the creation. These questions and their answers have baffled peoples' minds over the millennia and remain the same even today.

How easy it would have been to get answers to these questions, if possible, to table the questions to the Designer! Of course, and because humans are earthly, doing this would be impossible or beyond human abilities. The only alternative is to seek the answers by examining what the Designer created: the earth and the extent of human capability, the universe. This book will explore all created to find the answer.

Among the Igbo people of Nigeria, the 'worth' of a man is measured by his achievements in the society - his unique personal achievements, not his father's or his mother's! The achievements generally begin with first getting married and start a family. Any man who has not reached this lowest level of the ladder of success was often, at least theoretically, not accorded men's full rights in society. In the village meetings or such gatherings, such men were not given 'full but half-cups of wine to recognize their statue or level in the village. Even achieving this level was not a piece of cake': the young man had to prove to his family and the would-be in-laws he was up to the task of having a family. In other words, he had to demonstrate he could take care of himself, a wife, children and maintain a family.

Various communities set their terms like a show of barns of yams, ownership of large numbers of animals like goats, cows, landed properties, etc. In each case, the young man was required to extend his wealth to his would-be in-laws. This requirement was how 'bride price' was born. This method of proving oneself ready for marriage persists in some of the Igbo communities even today.

Currently, that method of assessing worthiness has taken a new form. Suppose a young man acquires an education, which would be an acceptable equivalence because education was a sure way to make money and live a decent life. In those days, obtaining a 'Standard 6' certificate ensured a permanent and lucrative job. If a young man is engaged in a profitable business, that would also be acceptable. Until very recently, no families allowed their daughters to marry men with no means of livelihood.

The young woman was expected to have gone through a family education – raised under the tutelage of a steady-headed mother. In those days, unlike these days, there were many such mothers. Such a woman was first a wife/comforter, then a mother. In the absence of the man, the wife knew to take over the management of the family. The terms wife and mother have deep meanings in those days in Igboland. Then a family was made up of a father, mother, and children.

Getting married and raising a family are placed on the last rung of the lather of success because this is what all living things do. A dog raises a family, so also a rooster and all animals, including plants. Therefore, there is nothing spectacular about raising a family. However, there is a complete description of a man's worth in his society – what he did to better himself and his community.

The killing of a lion, tiger, or any of the wild beasts was a spectacular feat for one man in the olden days. People told and retold the story far and wide. That would enhance the village's respect and status: a community that produced men of that caliber! Some men traveled far and wide and brought home some new ideas that improved their people's lives, like new farming methods that yielded more bumper harvests or new ways to build more modern houses.

There was a story of a man who, during his travels, met, fell in love, and married from a very wealthy family. The rich in-laws extended their friendship by building roads that connected his village to others.

Likewise, since we cannot see God, neither can we talk to Him; it makes sense to study or observe nature around us, the very handiwork of God, to find the answer to why the earth's unacceptable state. How could God create the world and furnish it with all the right things for all living things to use and enjoy, sit by, and watch all the unhappiness on this earth? That sounded very odd. Or are we missing something or wrongly taught by the teachers of Religion? My fascination with and the quest for nature started early in my life. Please read on to peep into my early days.

The good old days! Can it ever be the "Bad old days"? Generally, no! We reminisce about the early days with nostalgia. Yes, many would like to relive them. Why? The answer is simple for why this feeling: a young person with no worries at all, only getting ready to join the adult world. It might take another six to ten years to get there. The world was good. Whenever the father was around, there was no feeling of any danger; secured from physical harm; the loving mother was there to cater to hunger and wellbeing. Children had plenty of pleasure times to run around and discover things and places, unaware of what was happening in the world around them – the unhappiness, the wars. Yes, that was the good old days.

I still remember that stage in my life. I remember the first time I ever wandered into a nearby 'forest' all alone, by myself, of course, for no reason. What I saw then as a forest is now a little garden, where my mother planted pepper, okra, different types of vegetables, and even yams. Like other grasses, some tiny plants, and some big trees like the palm trees, I saw these and more. Bananas, oranges, mangoes, peas, raffia, cola nut trees, etc., were 'planted' nearby.

Of course, it took some years for me to master the names of these 'wonders.' Looking back, I think what struck me most was not only the sheer number and variety of these things, but now, suddenly, I became aware of them. I remember the first time my mother took me to visit my maternal uncle and his family. On the way, I saw the same types of things I saw in the forest: again, various grasses, small and large trees, etc.

Later in life, I learned these existed everywhere, including in other 'countries' around the 'world.' My world was my village. My little mind was wondering how many of these things lived. Besides, small, and big forests existed, and animals of different sizes and behaviors inhabited the woods. Animals, plants, and trees must be uncountable!

I never imagined how the woods came to be or who put the animals in them.

And before I encountered a snake, I had already developed a fear for it. I never understood how the anxiety started. Then, one day I got a lecture from my elder brother on what to do when visiting my friends from another family to ward off any animal, particularly the feared snake.

In those days, it was trackways that connected families and even connected villages to the villages.

He told me, "Little brother, you must make noise with your feet; stamp the ground with your feet or drag them. Better still, carry a small stick and bit both sides of the pathway as you go. Any of these actions will warn any animal that a human being was passing by." He assured me that all types of animals were afraid of humans, including minor children like me. That was comforting.

Since then, I have been interested in nature, in all things around me. I still remember the first time I saw the inside of a palm kernel—you had to crack the hard shell with a stone to expose the softer edible nuts. My fascination was why was the edible nut hidden inside a hard surface. I also saw the inside of papaya, orange, pumpkin fruit, lime fruits, cucumber, etc. (fig. 2), and later in life, the inside of an egg, animals like goats, rabbits.

These things were only different in their natures but the same in any other aspect. No matter the kind: a baby in the womb, a chick, a seed, name it: all connected to their food/ mother through the "umbilical" cord.

The arrangement of seeds in some fruits like Cantaloupe, Red bell pepper, Kiwi, Figs, Green pepper, Lemon, Tomato was wonderous to a child. I could see a perfect design, order, and consistent natural law at play – each seed connected to its food through the 'umbilical' cord.

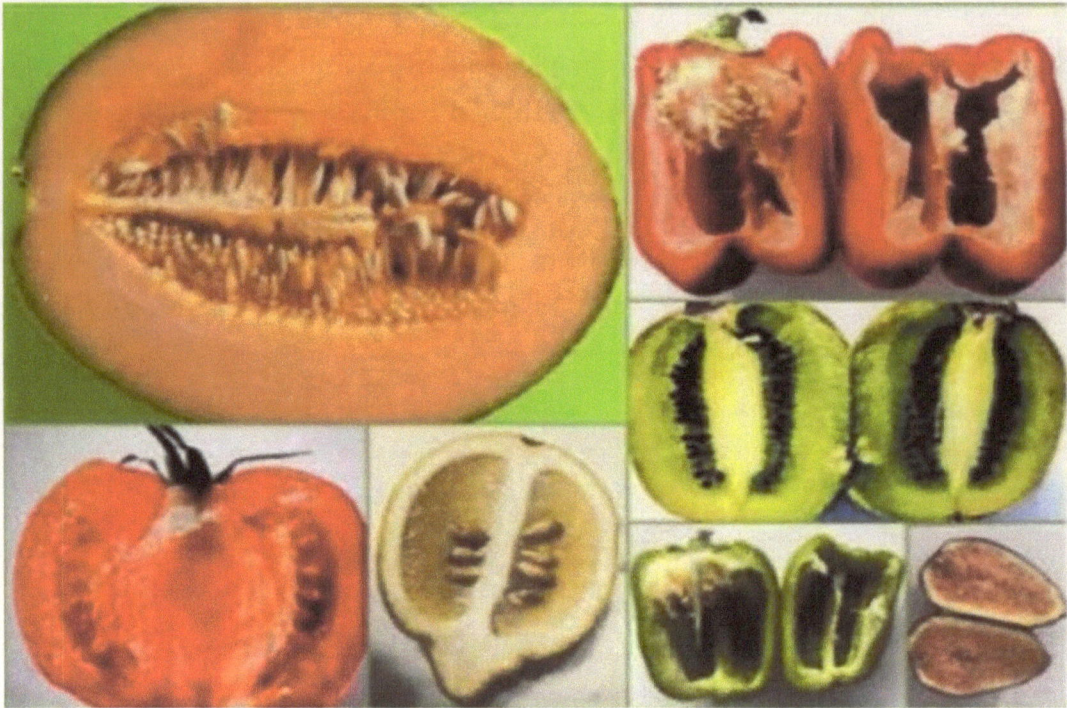

Fig. 2: Cross-section of: Clockwise, Cantaloupe, Red bell pepper, Kiwi, Figs, Green pepper, Lemon, Tomato, showing all the seeds connected to their food through the 'umbilical' cords before separation.

The connection of food to 'seeds' in fruits (plants) and 'eggs' in animals are similar: the 'umbilical cord' connects the 'seed' to the 'mother' for its food. As living things, the seeds and eggs must eat. But they cannot provide food for themselves. Therefore, the All-Knowing Creator attached their food to them before separation from their 'mothers.' We can understand why animal parents must feed their babies because they cannot provide for themselves.

The Creator's universal law is evident here: all living things must eat. First, it shows the importance the Creator placed on eating food by all living things. And He made food readily available. Secondly, all adult living things "must go" for their food and provided the means to acquire the food. Why? Why is the food not attached to them as in 'seeds and eggs'?

Another wonder was the Sun. When I was a child, the only means we had to ward off cold was either by the Sun's heat or small fires from burning wood. Then I came to understand the dependence of all living things on the Sun. Nothing grew downwards; all must face the Sun.

Only one Sun! Initially, I thought every village had its Sun. But recently, some scientists predicted that the Sun would continue to produce its heat for, among others, energizing all living things for the next seven trillion years. Seven trillion years! It will burn forever (beyond imaginable human existence), in other words.

The Solar System   The Universe

Fig. 3: The Solar System & The Universe

Our Solar System and the Milky Way Galaxy are approximately eight billion miles and one hundred thousand light-years across, respectively. The Galaxy is among the smallest galaxies with more than one hundred billion stars. The head-spinning part of these facts is that thousands of Galaxies form 'part' of the Universe.

It is reasonable to conclude that the Milky Way Galaxy is like a dot when the size compares to that of the Universe. But, of course, the Creator of these heavenly bodies can count the number of and measure the distances among His creations, a feat possible only to Him, the Almighty God. Incredible!

From these facts alone, we can conclude that the Creator is a spirit with infinite knowledge and power incomprehensible to humans. (For verification and more info: google the Solar System and the Universe).

I remember my first time in an airplane, about thirty-three thousand feet above the earth. There were no houses, forests, or animals. There, birds were absent. You can see miles and miles, as far as the eyes could see. There, nothing on the ground was visible to the necked eyes. It was awesome. You could not help appreciating the majesty and power of the Maker.

With maturity and more education, I came to understand that I was only in our earth's atmosphere; that there were other heavenly bodies like our earth, about nine of them forming our Solar System. And that from our planet to the mighty Sun was about ninety- three million miles.

Talking of a million miles, I used to think that the longest distance was from my house to the farther-most marketplace in my village. Now I know it is approximately five miles away that took about forty-five to sixty minutes' walk to reach. Then I learned that a journey from my village to Lagos, Nigeria's capital City then, took about a day and a whopping seven hundred miles! It was then nerve-racking to even think of seven hundred miles. Much later in life, I realized what seven hundred miles meant—a day's journey in a bus to Lagos. My knowledge of this distance was still an imagination: I knew what a bus was like, what a day meant, but I had not experienced the journey itself or been to Lagos.

In elementary school, we manipulated numbers: we could add, subtract, divide, or multiply numbers to get seven hundred. But what was seven hundred miles? When I traveled to Europe in the seventies on my way to America, I experienced distances in thousands of miles. In my mind, I had a picture of six thousand miles. The only way my mind could picture what a million miles was like was through association: How many times would I go to and come back from America to cover a million miles. The calculation is the best possible estimate because no two horizontal points on this earth are one million miles apart.

Then I learned that the planet Mercury is approximately thirty-six million miles from the Sun, our earth, ninety—three million, and the one-time planet, Pluto, about three and a half billion miles. A billion miles! While I struggled to understand what one million miles were, I needed to understand what one billion miles were. To complete my wonder and confusion, I learned that our Solar System was approximately eight billion miles across.

Then came the knowledge that distances were measured in lightyears – the distance light travels in one year at its speed – six trillion miles or six million-million miles. This information did not make too much sense because there was no picture of it in my mind, no reference point. The big question was, "Was there in existence a place it would take one lightyear (or whatever) to reach? "Yes," I learned later. The distance to the nearest star to our Sun, the Proxima Centauri, is approximately four lightyears. *(These astronomical distances can be verified by simply googling them).*

It takes about fifteen hours to travel from Nigeria to the US via Europe - a few thousand miles. The US Path Finder took about seven months at a speed of approximately seventy-three hundred miles per hour to cover five hundred three million miles. That was the distance to reach our next-door neighbor planet, Mars. Can you even imagine planning a journey to the nearest Star, the Proxima Centauri?

The shocker came when I learned that our Galaxy, the Milky Way, belonged to one of the smallest Galaxies. Note: "one of the smallest galaxies!" The cross-section of this 'small' Milky Way is one hundred thousand lightyears. To cross this tiny Galaxy will require a journey of one hundred thousand years at the speed of light. The most exciting part is that thousands and thousands of these galaxies, both large and small, form part of the Universe.

Scientists believe that there are other Galaxies they may yet discover. How long is the cross-section of this Universe, and how many light years will it take to cross? The answers are good left to the imagination. Not all minds can handle such incredible numbers and distances. Now, we know that our Universe is visible; but some others are not. I will keep the discourse visible. Where will any ordinary human being start to explain the invisible? Please bear with me.

But do not lose sight of what brought us to this point. Remember my explanation that in Igbo land, it's the man's achievements that determine his worth. The only way we will begin to understand the power, knowledge, and majesty of the Creator of all the things around us is to examine what He created.

So far, we have discussed the incredible and nerve-racking numbers of some of the heavenly bodies and brain-cracking distances among them. We know the Maker of these things can count them merely as a child would his toys and measure the incredible distances needing no help. How do you feel about the power and knowledge of the Designer of these heavenly bodies and distances so far? Is it possible the Designer can forget anything or that our earth's problems are too much for Him to solve? Or did He design it to be so?

We see many human-made technologies – household items like kitchen utensils, furniture, electronics like radios, television sets, computers, vehicles like cars, motorcycles, bicycles, traffic systems, high-rise buildings, skyscrapers, etc. We admire these things – the ingenuity of the designers.

Most of these objects took years and many people to design and build them. Then think of our earth, one of the planets in the solar system, and the universe. Scientists say our solar system is almost invisible from the Milky Way Galaxy's edge because of its distance. Besides, the solar system's planets revolve around their imaginary axis, and at the same time, orbit around the Sun. Also, each of these trillions of heavenly bodies was hanging on its own 'without support.' That knowledge was the greatest wonder to me. Again, think about these incredible distances and numbers, and then imagine the Creator's knowledge and power! Is your brain not spinning? Mine is!

Incredible! Scientists, in this context, will talk about gravity. What was the gravity made of: from what materials and by whom? So also, are the billion trillion of such bodies forming part of the universe? There has never been a record of any of these planets colliding with one another. Think of these for a moment!

Compared to the size of our planet earth, a car is like a minuscule speck of dust, if at all visible. You are aware of the number of accidents we have designing and operating our vehicles, locomotives, planes, and ships. If you take those into considering, it is easy to understand why humans began to reason that a non-human created our earth and the universe over time. We call the Creator of those incredible things God in the English language.

What is the lesson learned here? First, the Designer, the Creator of these things – the universe—must be a spirit. He can't be understood as a regular spirit but with infinite knowledge, power, glory, and majesty. Can anyone imagine a human being considering crossing our Milky Way Galaxy, one of the 'small' galaxies, a journey of many thousand lightyears? It would then be total insanity for a human being to plan a trip to cross the Universe. Take a moment to think of the Maker, the Creator of the Universe.

If a person or group of people built houses or cars or anything, they could always care for these things. They can, for example, see them, count them, repair, and maintain them. If any is missing or not working right, they will know. Then think of the Creator of the Universe! He is the same Creator that designed all the billions of heavenly bodies and kept them hanging and moving in space without supports and with no accidents ever recorded! He knows where each is at any time.

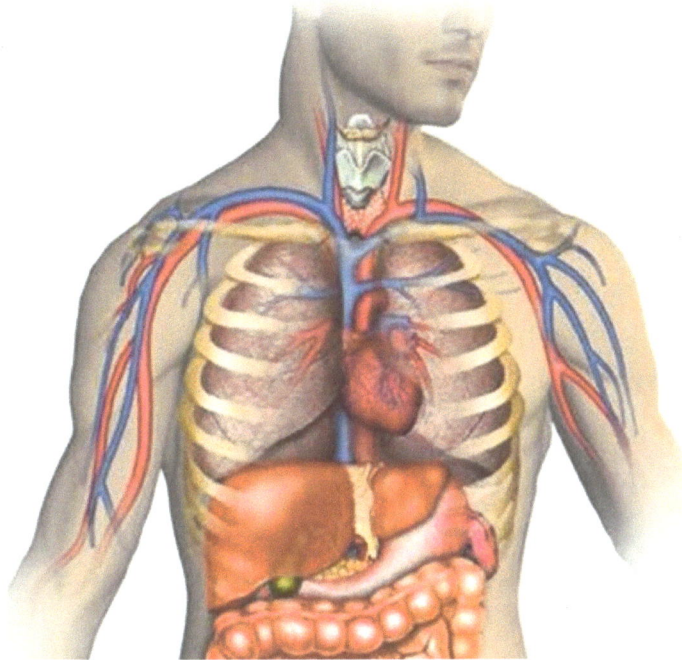

Fig. 4: Human Body Systems

It is simply miraculous that the human structure (and other animals) started life as a single cellular organism before the Creator-induced cell divisions began. The Creator had programmed the cells to acquire different characteristics as they divide and multiply to suit the work ahead. Hence, the cells of the eyes, nose, ears, bones, skin, flesh, hair, etc., are different.

Imagine the millions of cells in the animals' (or in any living thing) body? The Creator provided for each cell, though in millions. Each organ knows its position in the animal body: the head knows to develop at the top of the human structure, not at the bottom, side, or other locations. The system is already in place before ever fertilization occurred. Imagine the thousands of cellular functions in your body while talking to somebody, sleeping, awake, eating, etc. Most of these activities go on without attracting any attention.

The design of the human body systems is beyond human abilities. Nevertheless, there is no doubt the Creator is of infinite knowledge and power. The Creator is God.

He is the same Designer of the complex and complicated human body (fig. 4.). Looking at the human body or any animal body from the perspective of a non-scientist, just an ordinary everyday Joe, one can see it is the most complex machine ever built, a feat beyond human duplication. Where can such an attempt begin? From where will the materials – the skin, the bones, and the flesh come? Synthetic materials, rubber, plastics, etc., cannot substitute or be equivalent to the Creator-designed flesh, bones, etc. What of the complex human brains and the connections to all the systems in the body? Can human-made computers replace this? No!

Just imagine what is going on inside a person while standing and talking to another person. While talking, all the senses are busy doing their jobs: you can see, hear, smell, feel, etc. The voice is automatically regulated to suit the situation – not too loud or low, and your sense of balance is activated, which keeps you standing without falling. The conversation can even be interrupted if either person's thoughts are distracted by a sensation that demands attention.

These are physical. Think of millions of other functions going on inside your body without your even noticing. Billions of cells make up the human/animal body (a significant number indeed). Blood is supplied to each cell through a network of intricate 'pipes,' just like the pipe connection for water in a modern house. Each cell! If these pipes/veins were connected end to end, the length would be in miles. Incredible! Even though these cells number in billions, none is forgotten. The Creator is aware and provided what each needed. Then think of the number of humans and animals on this Earth?

The above discussion is no lesson in biology or science. We all know there is blood in every part of human and animal cells. Some people go further to understand the composition and the functions of the blood. These people know that without the blood supplying nutrients from the food we eat and oxygen to the body's building blocks, the cells will die. But we rarely associate these miracles with the Creator, God. I'm implying here that we generally do not see the gift – the creation and design of these things and enquire from where He obtained the materials He used? Next Time you see blood, take a good look at it. What you will be looking at is alive with many functional abilities.

Chapter 3

ANOTHER SET OF WONDERS

It is dumbfounding the incredible numbers and distances that exist, including designing and maintaining all living things and systems that keep them functional. What are common everyday occurrences seen in or around us and among plants? Try it again: cut open a melon fruit or green bean pod, pomegranate fruit, guava fruit, orange, or other fruits. Try and examine what you see inside the fruit: the arrangement and order. Notice how the seeds connect to their food through the "umbilical" cord. All, no exception! Miraculous!

Eating food must be one of the functions of all plants. All plants have seeds. All seeds are attached to their food to keep them alive. Each source will eventually germinate, grow, produce its seed, and finally, replace and continue its parents' functions.' And the cycle repeats. *(see fig. 6)* Is it not evident that the Creator's intention must be for plants to exist here on earth forever? These plants are designed to produce food, medicines, and other things humans and animals need. So, animals, including humans, are meant to exist here on earth forever.

It is easy to understand the main functions of all plants then: each group will exist here on earth forever by eating the food already provided by the Designer and reproducing themselves through their seeds; produce food and nutrients for all living things. Because seeds can't go about finding their food, the "umbilical" cord is provided to each 'seed'.

to connect it to its already-provided food – through their 'mother.' But when separated from the plant, the food is already packed with the seed. All these events take place in a perpetual cycle.

Appreciate the ingenuity and the wisdom of the Creator! Only a Spirit of infinite knowledge and power can accomplish this feat. He is the All-Knowing Creator, God.

In animals, the arrangement is different: In animals that give live birth, the 'baby' (fetus) gets its food through the umbilical cord. But the parents provide the food and feed the baby after birth. Animals' eggs get their nourishment through the umbilical cord while in the 'womb.' But outside the 'womb,' the egg contains its food. And the parents provide the food after hatching until ready to provide and feed itself. Because we do see these phenomena regularly, we hardly recognize and appreciate them, not to talk of respecting and admiring the ingenuity of the Designer.

Please, I suggest you take a second look at the arrangement and the connections of the pulps and seeds of an orange fruit or any fruit. You will marvel at and praise the wisdom of the Designer. Do we still need more proof as to the knowledge and power or intentions of the Designer?

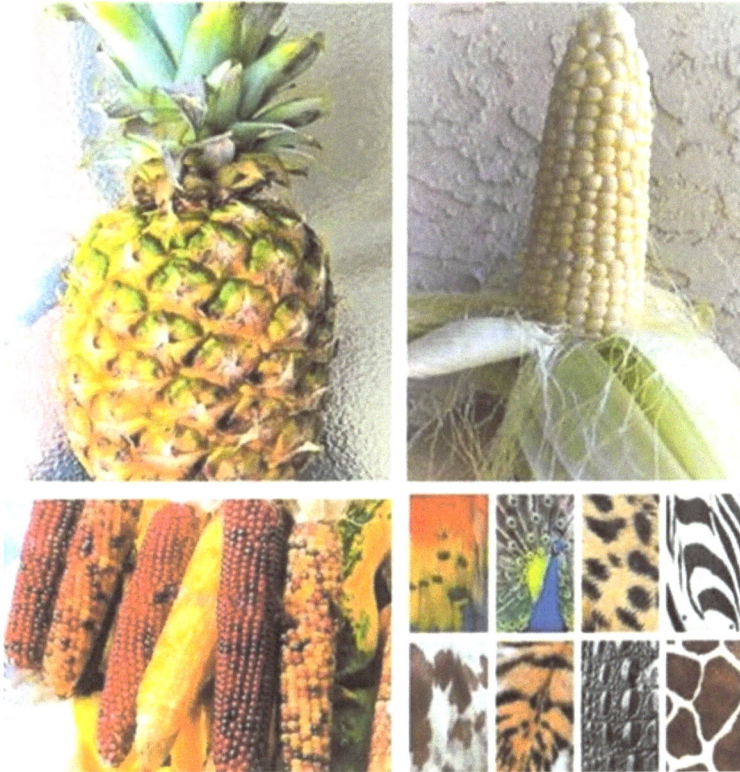

Fig. 5: (a) Pineapple, ears of corn (b) Animal Skins

Examine the pineapple, ears of corn, and the animal skins shown or any kind in your orbit. (The animal Skins – from top clockwise: Parrot, Peacock, Bobcat, Zebra, Giraffe, Alligator, Tiger, Cow). Then, take a second look at fig. 8 – a flower collage. The Designs and colors are simply miraculous! On this earth, the numbers of the Creator-designed colors and designs approximate infinity.

What might be the Creator's reasons for adding these beautiful designs and colors in His creations?

Take a second look at all His creations around you – humans, animals, trees and plants, the rolling hills and vegetation, and even the food we eat. Beauty! Not only did He provide the food, the Creator, God, also made sure He catered for and appealed to the individual tastes of even His 'picky' children. He is God of beauty and love. He does not give but in full measures. The provisions and care teach humans about love and not giving in half measures. There is no doubt that His original Blueprint for the earth was a Paradise Earth.

All my life, I have known and eaten corn. One day I walked into my little garden to check my corn stage in terms of maturity and harvested for eating. Then I plucked and shelled one. I felt like the experience removed a veil from my eyes. The design of the seeds struck me speechless. Please repeat what I have described. Incredibly, my corn turned into a piece of artwork. Look at the rows of kernels. Incredible! While carrying food and nutrients for living things, the seed has the usual connection to its food.

Then, out of curiosity, I started examining different types of corn whenever and wherever the opportunity presented itself. The result was the same. Also, I realized the corn came in various sizes, tastes, and colors. Etc. When you take a second look and experience what I am describing, the obvious question might be, "What was the intention of the designer in using such designs and providing them such beautiful and different colors?" I will leave this question for later discussion.

Then, my curiosity went wild and overflowed to other plants: any plant I encountered, I had to give it a second look—small plants – they come in different sizes, shapes, types, colors, different types of leaves, tastes, and even how they protect themselves to ensure continuity of their kinds; large plants – the same characteristics.

My curiosity again went wild and carried me from examining plants to examining animals. *(See fig. 5b)*. The result was the same – infinite variety in everything about them. After experiencing this, I came away with few conclusions: All living things must eat. How so? The food is in great quantity even to the 'seeds' through the 'umbilical cord.' And the urge to eat the food is always there.

Also, all living things must have 'seeds,' and a means to protect themselves. Just by observation, it was clear that the Designer intended that all living things will exist forever. The seeds will eventually germinate, grow, produce their seeds, replace, and continue the 'parents' work for plants. For animals, the 'baby' will be born or hatched, develop, have their babies,' replace and continue the parents' work.

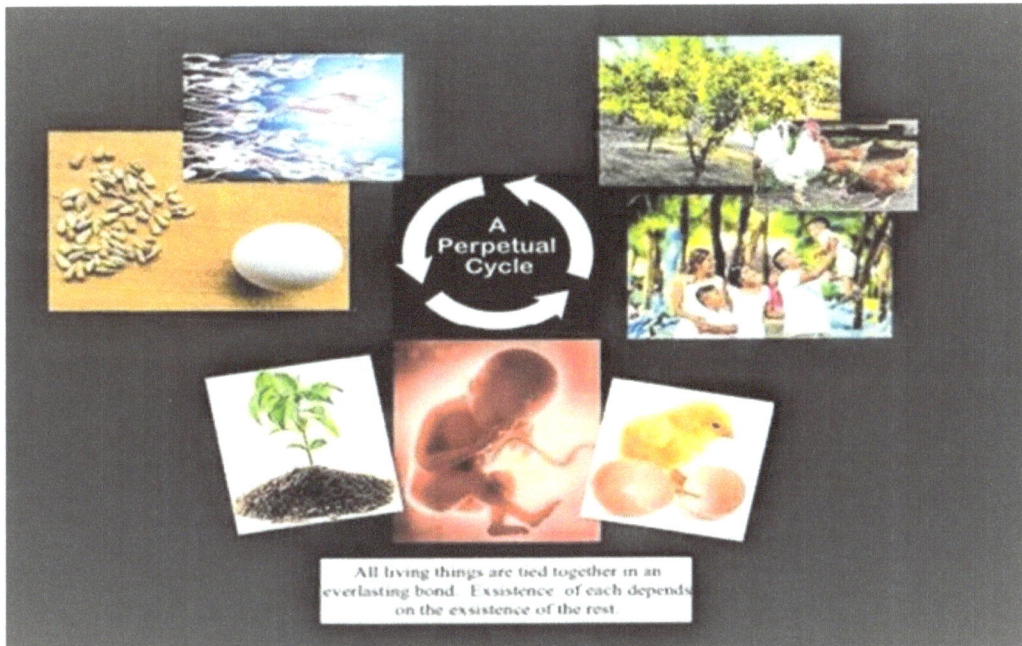

Fig. 6: The Perpetual Cycle

The cycle starts with the seed or a fertilized egg. Every plant has seeds or a way to reproduce itself. Animals have ways to reproduce themselves – through their fertilized eggs.

By the time, the 'parents' grow old and die, they already have their 'seeds,'; the seeds are by now 'adults' and ready to produce their own 'seeds.' Inevitably, they take over from their parents. Then the cycle starts again and repeats forever.

Thus, the Creator of infinite wisdom and power set the order. Obviously, He intends to have all His earthly creatures: plants and animals, to exist forever on this earth. He furnished them with all their needs.

See fig. 6 – the 'Perpetual Cycle.' To produce animals' and plants' food, water, air, sunlight, and good soil are needed. Since living things need their food to live forever, it is inevitable that the air, water, sunlight, and good soil needed to make their food, will exist forever.

There is no doubt these will exist forever. Since the 'seeds' are incapable of providing or going for their food, the Maker attached their food. For the 'grown-ups' – plants or animals – they must 'go' get their food. How? He provided the food, and the urge to eat them is in the system. But the food was not attached to the adults as the 'babies.' Why? Indeed, the Designer needed them to, among others, 'move' to 'get' the food. The means for doing this was available, proving it was the intention of the Maker.

Since plants generally can't move, though some float in water or air – how do they go about getting the food and feeding themselves now they are no more seeds that the Maker attached their food to them? They must live forever and, as such, must 'eat.' The Creator was aware of this 'deficiency' and has directed the Ns factor to instruct them on how and when to grow and send out their roots in search of food items and have designed photosynthesis to prepare the food. Wonderful!

There are too many differences among plants of the same kind. You can imagine the discrepancies among plants of various sorts and their root systems. Examine the plants' root systems, their types, shapes, sizes, and colors of the flowers. There are only two words to describe these phenomena: infinity and variety. I will expand on these two later.

Then my attention turned to examining other types of living creatures, the animals, in more detail. The differences were more astounding. Unlike most plants, the animals have means of moving about, mainly searching for their food or other activities related to food. A baby in the womb can't go looking for food. Neither can plant seeds. So, the ever- knowing Designer has their food attached to them. Yes, the Creator did not connect the food to the adults of both plants and animals. But instead, He provided them with means to move about to, among others, get their food. His intention must be that all 'adults' must 'go and get' their food. Why?

Some animals walk on two legs (like the humans); some on fours (like the cows, lions); some don›t need legs but wings (birds like the Eagles) to fly. Some don›t have legs or wings but fins to swim like the fishes; some still don›t have any of these parts because they can only crawl or float (snakes, amoeba). As for the movement of trees and plants to get their food, they have their roots and leaves grown in no time, searching for food items (the roots) and making the food (leaves).

As many methods, animals move as the number of animal groups on this earth is almost countless. When one examines their feeding mode and types of food, they eat again; the list goes on for what seems like forever. When their sizes, shapes, styles, colors, tastes, and how they protect themselves from predators to ensure continuity of their kinds, the list goes on indefinitely.

Another variety is these animals' strengths and power: it varies from the weakest (like the humans) to the most powerful (like the lion or the great white shark). It is instructive to note here that the Creator subdued many animals and plants for human beings. Also, the Creator blessed humans with the ability to develop intelligence.

Humans can deal with all the wild beasts that would have otherwise killed off humans with these two advantages. Human beings have few predators. All animals have a natural fear for human beings. Therefore, cows, horses, elephants, etc., that are about six to ten times the size and strength of humans are obedient to and can be trained to serve humans. They also provide food, including clothing, to humans.

Because of the possession of intelligence—humans are the only type of animals that have improved themselves. First, they can make decisions based on the stimuli interfacing with their senses in their environments. Secondly, humans can take care of themselves and all the rest of God's creations. And the Creator gave humans dominion over the whole earth because of the possession of the brain—intelligence.

Humans' fourth job is to take care of and protect the earth and everything on and in it. We know this is true because of developing a conscience and the two gifts mentioned above. A man has no choice but to do as the Ns factor dictates. Other types of animals possess instinct, which is only a part of the Ns factor.

From all these, it is reasonable to believe the Creator requires all living things to feed and protect themselves and maintain their kinds through procreation. How to do these are written in the Ns factor: both intelligent creatures, mainly the humans and the intuitive types of animals and plants—must feed themselves, protect themselves from predators, and through 'reproduction,' maintain continuity of their kinds here on earth. Following the Designer's infinity pattern, they perform these functions in an infinite number of different ways.

At the beginning of human existence here on earth, people wandered around searching for food because the Ns factor had instructed them that, to live, they must eat. The food was intentionally placed everywhere by the Maker. It is sensible to believe that the Creator 'commanded' all adult living things to go in search of their food. The food was available, and the means to get them provided by the Designer. The only requirement is, "you must go and find them." Till today, all living things are engaged in search of their food.

As a personal note: many people claim they have the freedom to do as they please. At best, the feeling's impact is minimal, but at worst, it isn't sincere. Yes, you can drive on the road's left side when the law is to drive on the right side. You know how far you can go? You have the freedom not to eat, sleep, take a bath, listen to anybody, etc.? Do you have the liberty to do as you please? NO!

At a point in this searching-for-food process, The Ns factor sets in, making the prehistoric peoples settle down in one place and produced their food. Wandering to look for food diminished considerably, and farming began. It is easy to understand that the cultivation methods and types of crops cultivated, including the types of animals domesticated, are of immense numbers. Settling down in one place and producing or 'buying' food to eat also fulfills the 'commandment,' "Go and get your food."

A coal miner, an office worker, a mom taking care of her household, the president of a country, a farmer – all are in obedience to the command, "Go and get your food." Mainly farmers and ranchers produce or 'get' the food these days in most countries. The rest of the adults must bargain with them to acquire food. "You must eat" is a natural commandment! I call it a natural commandment because our bodies have the urges that make us do so.

Since plants are stationary, the Creator equipped them with various root systems to do the food-gathering for them. Because of the Ns factor, the plant knows when to put out roots and send them to look for the food materials, essentially: minerals, water, etc. They naturally transport the food materials through the stem to the leaves, where, under sunlight, using carbon dioxide and chlorophyll, the food is manufactured in a process called photosynthesis.

It is, therefore, essential to note here the importance of sunshine and water. Nothing can live without these two elements. These are there for taking to all living things and enough for all. It is impossible to count sunshine or water. Again, the words "infinity" and "variety" come to mind. Think of how many different types of 'lower' animals on this earth and imagine the infinite ways they go about their food. Take, for example, lions, elephants, birds, worms, ameba, flies, etc. These animals belong to infinite different groups. When the types of food these other groups eat, or the different ways they go about finding their food, the number is mind-burgling.

Many animals are vegetarian, while many others are carnivorous; some are both carnivorous and vegetarian. Some insects and worms feed on animal blood. Many worms live in animal flesh, skin, guts, etc. In contrast, different types of insects and worms live in the earth. Even some bacteria live in the air. The Creator provides food for these creatures where they live. Incredible is the sheer number!

Could there be any other reason why the Designer required all "adult" living things 'to go in search of their food and didn't merely attach the food on them as He did with the "seeds"? An educated guess would be, He also wanted them to move around and exercise their bodies in the process, communicate and be neighbors with their kids, and help one another in the process?

As detailed later, the Maker did not give all knowledge or provided all needs to one living thing or a group of living things. Therefore, we must 'go' and get whatever information or material needs from other living things. The certainty for seeking assistance is also why no creature lives in isolation but communities. You can observe that no living thing, particularly animals, is ever at rest or in isolation except during the appointed time to rest. They are always in search of food or other needs.

Even in the invention, discovery, authoring, exploration, etc., all are achieved by many people›s synergy, i.e., using many people›s combined efforts. It is of the Creator›s design. Humans can›t see the Creator; neither can they have direct or physical contact with Him. But many have shown that they were sent or commissioned by Him to accomplish some tasks here on earth through their work. Then, the only way to learn about Him is by studying His creations in detail.

What is evident from the 'infinite variety' make-up in His creation is clear: No two things are alike or made the same. Each creature has a unique identity, even if from the same species, and it's for a specific purpose, known only to the Creator. Therefore, we should not be alarmed when we see differences in sizes, behavior, character, intelligence, etc., in all living things.

How can earthly beings sit in judgment of or evaluate or amend or correct what God created? Such an attempt would be childish and ridiculous. No living thing, particularly humans, should say, 'This earth thing is good or bad.' No person can say so because no one knows the intentions of the Creator.

Rest is not just for resting's sake. The Creator knew the bodies and the systems of living things; mainly, the animals would get tired and sore after a day's work of looking for food. The rest period He established to provide the body a time to rejuvenate and rebuild itself. No matter the quality of food you eat or the type of medicine you take, none will function as designed without rest. Even today, whether you work in the office or as a construction worker, a mom taking care of the family and children, the leader of a country, all need rest. So, the Designer decreed. There are no exceptions.

The Creator had a plan to ensure that all living things got their rest: some He equipped with eyes that could not see at night. So, they had to do all their work (food gathering) by day when they could see. For animals that could see at night, they do their work during the night and rest during the day. Those animals that can see both at night and day take their rest either during the day or night. What of the other types of living things – the plants? At night, they can't work, i.e., go in search of their food. Sunlight is not available at night. Yes, plants take rest at night.

All living things must rest was decreed by the Creator. But why so? I explained one reason above. And since nobody can understand the Maker's will, a second reason is an educated guess—He was protecting the different types of living things. Would not the field be overcrowded with all the living things looking for food all at the same time? Think about that! Also, because He is God of infinite variety. He does not limit Himself. You can see the Creator left nothing out; or forgot anything.

Yes! Because the Designer is the ever-knowing God. The takeaway here is simple: The provision of all these – food, rest, comfort, security, etc. to His creatures with nothing left out, shows He is of infinite knowledge and "cares and loves" His creatures. We should learn from the Creator, our' parent', that we can't provide to our neighbors in half but adequate measures.

'Resting must follow work' is an order. We can deduce from this that after creating all the needs of all living things, The Creator must have established His rules to guide behaviors and consequences for transgression. For example, lions, though powerful animals, are restricted to certain parts of the world. They are not allowed to step outside their boundaries without severe consequences, including starvation and death. Some humans' activities, like removing animals or plants from their natural habitats and relocating them in alien environments, jeopardize this natural order making humans wanting in equity, love, justice, etc., set by the Creator. The consequence is the never-ending misery on this earth.

Chapter 4

INFINITE VARIETY

I had wondered why the Infinite Variety in creation. The answer came quickly: Suppose all humans are of the same size, complexion, either only male or female, the exact height; what of having only one type of plant producing only one kind of fruit; suppose we have only one weather pattern: either cold or warm; etc., the earth would not be enjoyable, very monotonous. The freedom of choice, ease of identification, etc., will be affected. How do you recognize who is who and what is what? It is easy to identify a person simply by hearing the voice. Likewise, animals can be distinguished when you hear noise or sound from them. These tell us that the Maker is of infinite wisdom. He willed total comfort for all His creatures.

We see infinite variety in all God created: for example, the number and the different types of flowers, animals, colors, etc., (see figs. 7 & 8). If the facial structures of all human beings are examined, few billions of them, both living and dead, the expression "infinite variety" appears again. Examine the facial structures of Chinese, Caucasians, black people, Indians, Japanese, etc., the same facts are evident. Take, for example, the facial forms of Nigerians. Nigeria or any other country is made up of different ethnic groups. Even among these sub-groups, the facial structures vary.

Go further down the line, and you have families. Members of each family have, among other features, different facial structures. Why? The design makes it possible and easy for a family member, nearby family members, or acquaintances to recognize who is who in the family. Nigerians are only a small group of black people. Nigeria is made up of millions of families. Then imagine the number of different facial structures on this earth. As "infinite variety" describes the facial structures, it also describes any other part of a human structure. Each human generally has ten fingers. But each person has a different fingerprint. Amazing! Another description of the phrase "infinite variety."

When skin complexion or color is mentioned, what comes to mind is black people, white people, or red people? If you compare two dark skins, each has different hues of blackness. You can see that this is true of red skins or white skins. Also, you think of infinite variety when the number of people on this earth, both living and dead, is considered. It is easy to agree that the same is true with animals. Many animal skins, including human skins, are covered with hair. Feathers or shells cover some animals, while the worms, snakes, etc. have none of those, as noted earlier.

Take a second look at any animal and appreciate God's designs on the skin. The skin here is the covering. The design is a significant characteristic that distinguishes one class of animals from another. *(see fig. 5b).* It is easy to pick out or recognize leather from a snake, lion, cow, etc. Even among these classifications, differences still exist. The same is also true of plants. Plants' leaves are generally green. But a second look reveals that no two 'greens' are the same. Incredible!!

Let us look at the characters, knowledge, behaviors, etc., of animals, for example, human beings. We can see that there are no two human beings that are the same in all respect. If five, ten, fifty, or even a hundred or more people, witnessed the same car crash. Each narration of the incident will be different. Yet despite the variances, the common factors allow us to recognize it as the same story. This natural difference is an example of infinite variety.

Animals can communicate with each other through sounds or signs. In this scenario, only humans can talk to one other. Many animals can only make sounds. The type of sound also will depend on the kind of animal. Making sounds, signs, or talk is designed to aid animals in fulfilling their functions here on earth. Some animals use sounds or gestures to, for example, warn each other of danger or announce the discovery of a food source. Only humans can talk to communicate with one another.

Just imagine the first Time one ancient human talked to another. Indeed, the hunter/ gatherer settlement in one place to produce their food triggered the formation of communities. Talking to each other was inevitable to enable them to live and foster harmony in their communities. What started as signs, gestures, and sounds over many years, developed into languages. Yes, societies must have created their languages through signs and sounds. With Time, people began to associate certain sounds with meaning certain things or actions in the community. Thus, they developed their language.

There were many communities, so also the number of languages. These communities or tribes developed their unique ways of living—producing and cooking food and doing different things in the community like dancing, singing, and making music. As this (culture) developed in one tribe, it was evolving in other tribes but at different rates and in different ways, depending on the environment. The environment played a significant role in determining each type of culture.

It is interesting to understand how the men end up being the hunters and cultivators of food and the women the caretakers of the homes and families, including the babies, children, and the elderly? How did the children come to help their parents with the family chores? The Creator had written the answers in their minds. Nobody tells a goat, chicken to take the young ones to the field for food and bring them back at the end of the day!

A man must provide for the family. He knew that. Going in search of food included hunting. Hunting involved strength, some danger. Also, there was danger moving from place to place in search of food. The man knew that nature equipped him better for the jobs that involved all types of risks without saying it. So, when he went out to hunt or do the heavy lifting on the farm, the 'wife' knew to take care of the family or help with nurturing and less strength-demanding chores.

There is another conclusion about 'infinite variety' – outliers. Let's examine the infinite variety again as God made them. Consider living things' sizes, colors, behaviors, types of food they eat, how they go about getting their food, names of living things, abilities, relationships, etc. In all these, each has its extreme. Take skin color; for example, some dark skins are so dark they are described as unusual. Also, about white skin – some are so white they are also described as unique. Examine relationships/desirability — ordinarily, males and females desire each other. We should expect to find some outliers, as is true with any creation. Again, it is seen some males choosing males or females desiring females.

There have been many arguments among humans about some of these extremes — homosexuality, for example. In the early stages, when people with different sexual identities started coming out of their closets, people advanced various reasons for the motivation for this type of lifestyle. Such people, they argued, lacked self-control with regards to promiscuity.

As a heterosexual, the thought of a man having sex with another man is unthinkable and repulsive. The same feeling is true among straight women about a woman having sex with another woman. It was later that some people understood that this type of behavior belonged to the extremes, or it is an outlier typical in all creations. Shades of blue color are infinite: The two extremes, the darkest and the lightest blue, are so different that many people will reject them as blue colors.

The question becomes, "Why is it that 'outliers' exist in all that God created? An educated guess (no human can decipher the will of the Creator) is that "It is the rarest among infinite variety – the way the Maker intended." Acceptance of or living with these outliers has caused a lot of unhappiness and grief among humans. Humans don't know what the Creator knows. We should accept that the outliers are the extremes of the Creator's infinite variety and remain confident that God is good and all that He created is for the good of all His creatures.

Chapter 5

THE CREATOR'S RECYCLING SYSTEM

Looking at all the natural phenomena and wonders and miracles we see every day and around us, nothing shows that there was a time when the Designer of all things engaged in verbal communications with humans or animals, and indeed, not with plants. Why would that be necessary anyway? From observing the creations, we know that the Creator is not human but a supreme spirit in all forms of existence: no human can create even a fly, not to talk of creating the universe and the unseen things. Spirits do not speak, eat, or have human earthly desires. Yet all Creator's creatures, except humans, comply with His laws. The Creator established the instructions and regulations through the Ns factor.

So, what is the Ns factor? Stated, it is the force of nature in all living things that makes them do all they do naturally without external influence: A little bird teaches its young how to fly; when you are hungry, you look for food; a tiny seed, in the ground, knows the time to grow roots and when to put out leaves and even knows when to send out the roots in search of food materials.

Check out the ultimate miracle – you. You started as a single cellular sperm. Nothing but the Ns factor will instruct it to grow and multiply on entering your mother's egg. Who taught it on when to grow tissues, with bone tissues covered by skin tissues? How did it know where to grow the head—at the top or bottom of the human structure? No external force but Ns factor that instructed it to hide the brain in the hard shell called the skull.

Yes, so it is. Our earth replenishes its resources from a perpetual divine recycling process without any human or external influence. There are no new creations. The

Creator does not create 'new' water or air or anything as the universe's design is complete. A new scientific discovery is indeed a discovery, not newly created. The object has been there but not discovered by humans.

Check out a bucket of dirty water from the laundry. We cannot drink or put it to any use. Really? Alternatively, some people may throw it out to water the lawn or garden or wash the driveway. In this case, people might assume that they will lose the bucket of water forever by so doing.

But no! In the garden, plants will absorb as much as possible while the living organisms will take their share. Some of the water will acquire the latent heat of evaporation and join the atmospheric water system. And upon condensation, the water will fall back to us in any of the forms of precipitation.

At this point, what remains of the dirty water is solid dirt that will separate and stay at the soil's surface. The final part of the dirty bucket of water is clean water that will join the groundwater system, available to living things from the streams, rivers, etc.

This process, part of which is the water cycle, continues to repeat itself forever. When living things die, nature does not discard their bodies but uses them to feed the living. Indeed, their remains do not go to waste. The Creator designed the system this way. And this is why the system happens with no prompting or external influence. We call this order the Ns factor.

Once I explained how Bees made honey, the sweet liquid many people love, to middle school-age children. I was trying to teach the Creator's giant Recycling System.

Honey, in a real sense, is the feces of bees. Some of the children were about to throw up when they understood from what honey was made. Bees' food comprises nectar from pollen, some honey, water (some people believe the water may be regular water and even urine), and many other materials the bees regarded as food. Is there urine in honey? Of course, no? Urine is a solvent. When ingested by the bee, it separates into its parts. The bee's system chooses the parts its body needs for food.

Even raw sewage from our toilets, bathtubs, kitchens, etc., is used to water fields, lawns, farms. Even solid wastes are used to fertilize farm crops and are used as animal feeds in our pens and poultry. Many animals (even humans) urinate as they wander or move about for whatever reason. The question is, what happens to these ‹waste› materials?

In the Creator›s system, these materials are not ‹wastes› but recyclable materials. Urine is a compound of different components. So are the liquid or solid wastes from our sewage. When one eats tomatoes, apples, vegetables, etc., grown using these ‹waste› materials, he/ she is not eating waste materials. Why?

First, the All-knowing Creator uses these so-called wastes in His giant recycling system to keep these crops available forever. Secondly, before these 'wastes' – solids or liquids – are used to feed the living, they are broken down into their components. Water in your bathtub after a bath you can't drink; it is 'dirty,' and it is now a waste. No! It is still useful.

The Creator, the God, sees it as an earth material and must stay on earth forever and be reused. The God-established process of separation will immediately go to work. Pure water will separate first through evaporation and later to condensation and finally comes back to us as precipitation, pure enough for use by living things. The oil part will separate to be used by living organisms. Dirt from the environmental dust and dead skin cells will separate to be used by living things. There might be some solids like the sand, which will join other sand. Nothing will disappear.

God, the Creator, has His way of preserving His creations, keeping them the same as He intended. There is trouble or consequence if humans try to change the nature of any of God's creations. Let me cite one example to explain this concept. If there are many years of cultivation or grazing or some form of abuse like the burning, a piece of land will begin to lose its nutrients and fail to support agricultural activities as it should. Will this render the piece of land useless and then be dogged up and thrown away? The land is an earth material that must remain here on earth.

What are the choices for farmers to remedy this kind of situation? As the Creator intended, the land should be left alone, fallow, for some years to allow the Creator's process of restoring such land to its original state to take its course. Describing the Creator's means to restore the land is unnecessary since this is a known fact.

Instead of humans following the Creator›s instruction will use human-made chemicals, believing they would achieve equal or better results. We know the type of crops produced using man›s method – food our bodies and animals reject. And consequently, humans incur the wrath of the Maker – inflammation, diseases, and unhappiness – for attempting to alter His design.

One may be tempted to argue that human intelligence, the brains, were given to humans by the Creator to invent things that can improve their lot here on earth. Humans invented airplanes, high-speed trains, the internet, clothes, guns of every type and strength, bombs and atomic weapons, and many other things that separated them from other animals.

Imagine humans without clothes or eating their food raw! The more important question should be, "With these new technologies, are humans happier, the earth more peaceful, and love overflowing everywhere as the Creator intended?

The answer is NO. Humans are incapable of inventing things that can bring happiness and order to the earth. The Creator has accomplished those already. But humans fail to reflect on and appreciate them. The problem is humans' inability or unwillingness to follow the Maker's prescriptions as written in His creations.

Humans are not allowed to alter or change the designs of the Creator. How can we be sure this is true? By observing the disastrous consequences of or retributions incurred by so doing. Has the invention of instruments of war, including the atomic bombs, brought happiness to the earth? NO! What of altering the composition of the soil we grow our food in, our seeds and grains, injecting growth hormones in our cows, chickens, and goats that produce the meat, milk, and eggs for our food. None of these has brought happiness to the earth. Humans are given the intellect and all they need to live like humans and take care of the earth.

Humans are even attempting to ‹create› animals, including human beings. Only the Creator can ‹create› because to create is to make ‹out of nothing.› Humans can only use materials already created to invent or ‹copy› from the «parent,› God. Scientists, copying from the Creator, can introduce human sperm to a human egg to produce fertilization in or outside the womb, in a process called ‹artificial insemination.› This fertilized egg can grow and develop normally in the mother›s womb and is delivered through the ‹birth canal.› The Creator has His reasons for requiring that the fertilized egg must develop in and nurtured by a human life womb and delivered through the birth canals. Humans now understand some reasons why the Creator required it.

Suppose this fertilized egg develops in and nurtured by an artificial womb; what type of human being will the baby grow to be? What kind of character; will it have human feelings? In other words, will it be a rational human? Would not this spell disaster?

Imagine what goes on in big halls or enclosed stadia, filled with people. In some cases, people serve food in such places. Take your mind to a beach on a typical day, particularly during holidays. Peoples will fill that part of the ocean or sea. The common denominator to happenings in all these places is the visible and invisible pollution in those places. Baring communicable diseases, organic pollution will be natural and always taken care of by the Creator's purification processes.

Tell me more! Please think of how many people will vomit into the waters because they accidentally swallowed the ocean or sea waters or pass unnoticed, some gaseous or liquid human wastes into the waters; some people will sneeze, cough, and blow their noses. Of course, many will talk, spreading saliva all around. The Creator knew and had His invisible workers already in place to take care of things.

In the halls and stadia, many will discharge some gaseous human waste maters. You see a miracle when people open the doors and windows, or people leave the waters – the Maker's invisible workers will move in and eat up what we termed wastes, restoring the previous conditions in these places.

Handling things this way is the Creator's method! But the 'modern' person will not listen to the Maker and will use non-biodegradable plastic pollutants (cups, straws, bags, and plates) to refresh or clean up the area. His way of cleaning the swimming pool is by pouring toxic chlorine into it. And to clean the halls, he will spray toxic chemicals like disinfectants and air fresheners. No! The use of these chemicals is not the Creator's prescription. We know the consequences of this disobedience. How can earthly and created humans think they can improve what the Creator fashioned or do better than the limitless and all-knowing God?

I have already mentioned that nature uses the dead to feed the living. The method is a natural process. People can see this process in different forms in human and animal interactions. Think of what happens to the carcass of a dead animal in the forest: many beasts will pounce on it to feed; vultures will descend on what remains to get a meal. Finally, ants will move in to cart away the tiny invisible parts. In the end, what remains of the animal would be the bones, too hard to eat, and only right to join the fossil world. You can guess correctly that this will be an addition to our future› crude oil› components. Nothing goes to waste.

A similar process takes place when a human being dies. Yes, the relatives bury the body. But what happens to the buried body? Other animals and even plants will feast on it!

A vehicle mechanic was boasting of his good fortune and was considering expanding his business, hoping he would be so lucky in the coming years. You might be curious to know how he came about his good fortune. That year there was prolonged inclement weather resulting in lots of vehicle accidents. As a result, he was overwhelmed with work. The effects of the bad weather were why he was thinking of expansion. A similar situation will occur if many deaths were requiring coffin makers to make lots of coffins. The business would boom.

Those few examples showed no waste in nature; there is some good in every situation. Because we are humans, earthly, we grieve and cry over adverse conditions. It isn't easy to look beyond that to see the good in every case. We should learn from the experience.

Chapter 6

OUR BEAUTIFUL EARTH

What else do living things need here on earth that was not already provided by the Creator? Our duties here on earth, for the nth time, are: Live and continue our kind. Humans have the added responsibility of protecting and taking care of our planet. The job was not to be a punishment. Humans should instead enjoy their work because it is of the All-Knowing Creator. Humans don't know what God knows. The day was designed for working while night, the rest period. During the night's sleep, the body rejuvenates. Nothing can substitute the night's rest. Besides providing the food from which the body gets sustenance and medicine, He also provided air, water, sunlight, and good fertile soil.

The Creator inspired how to make music and fashion musical instruments, including furnishing materials required to make them. Remarkably, people produce songs and music in all languages. One does not have to understand a language to enjoy the music composed in the language. Humans, including other animals, can enjoy songs and music from any culture.

At the end of the workday, families assembled to listen to music, watched dancing, listened to stories, or enjoyed the company of one another. On starry nights, family members sit out to enjoy the wondrous sights of the stars and the heavens. We will not forget the early morning songs of birds. In spring, the eyes feasted on the plush greenery; rolling hills and the mountains were green and full of life.

Fig. 7: The Apple Collage

This collage shows apples of various varieties, though some cultivated. They range from: red, green, pink, yellow to multicolor. Please note that they are also of different sizes, shapes, etc. But they are all apples. Just imagine how many apples exist on this earth! One can see this phenomenon in all God's creations – nothing is one type, color, shape, etc., and the number approximates infinity.

The Creator only is one.

The infinite variety in creation shows that the Creator, God, loves and cares for His creatures.

Suppose He created one type of apple or one type of anything with the same size, color, etc.; how will the 'picky' humans and animals enjoy their stay here on earth? The Creator is All-Knowing and All-Caring.

All the equatorial regions of the world enjoy those gifts of God throughout the year. The life-giving rain is in plentiful supply. Animals of different kinds, flowers of various types, forests, and rolling hills, are full of life. All animals are busy fattening themselves or preparing for the next season as the Creator had ordained. Food is plentiful, and there is joy in every heart. The time is right for producing the next generation.

Generally, spring is the blooming season in temperate regions. Flowers of different shapes, types, and sizes appear to please the eyes and gladden the heart of humans and other animals. God, the Maker, even made our eyes and those of many animals to be able to recognize and enjoy colors. Even the air we breathe in is occasionally 'washed' when it rains. In all these, human beings understood that nobody tied them to a place. In other words, people can move from place to place and can live in any area of their choice, in their environment, to enjoy God's blessings. Can it be better? NO! No human can do better than the Creator.

What do we learn from God's gifts and bounty? First, we should know that these gifts are limitless and should enjoy them. Secondly, we should embrace our duties here on earth through living in communities and taking care of each other. Thirdly, we should realize that all of God's bounties belong to Him and are for us to share equitably and use to aid us in our journeys here on earth. Finally, we should imitate the Creator in how He loves and takes care of His creatures to take care of ourselves and the rest of His creations.

He provided the food with their life-saving nutrients. To some of His creatures, He attached the food on them – they don't have to go, or rather, they are unable to go around looking for the food.

To others, He instructed them on how to get all the food they needed. Besides, he equipped the animals with the brains to help them navigate their way here and take care of the earth. The brain is like our earth machine, the computer. The computer can be helpful only when you input some information into it. The Creator God knew that the brain was an information-dependent machine. Hence, He created the senses.

He gave most of His animal creatures, particularly humans, five senses: sight, test, feeling, smell, and hearing. We should not forget the balance. Think of how humans or any animal would feel without the gift of the eyes, for example. How would the animals see and enjoy the good things around them – like the family members, the sky, the trees, the beautiful flowers, or even something that could harm them?

Think of how the sight of your loving spouse, children, or friend coming to welcome you gladdened your heart as the day's problems encountered in your workplace disappeared. Have you watched school children playing; what of two lovers, two puppies, or even Dolphins, playing or teaching their young ones? Beautiful experiences, right? These are made possible because of the gift of sight, a blessing from the Creator.

What of the sense of smell? Do you remember how your mouth watered when you passed the fast-food place simply because the aroma oozing out from it aroused your sense of smell? I still remember how happy I felt by smelling the aroma coming from my mother's kitchen when I arrived home from school. Do you know some people are turned on when they perceive the smell or scent coming from the clothes of their loved ones in their absence?

The impulses for the sense of feeling come from different sources and not only through touching or physical contact. You can feel using your sense of sight: When you see a group of people dancing to a piece of melodic music, enjoying themselves, one feels to join them. You know the feeling when children or even adults going through some pains or difficulties get hugged. Even some people ask, "How can any person love a child he can't even touch?" Touching or physical contact, when effectively use, can work wonders. The senses of hearing and taste also have unique influences on our lives.

Consider a deaf person passing a railway crossing or taking a walk on our busy roads unguided. What a disaster? If we don't have a sense of taste, how will anybody know what is bitter or how can the test buds feast on sweet things? The Creator God knew that humans, including other animals, couldn't have fulfilled lives here on earth without these senses.

Many argue that only humans have real brains, but other animals have instincts. This school of thought buttresses its position by showing that animals have never improved themselves – but only humans. I will ask these critics to visit a zoo and watch a gorilla tend to her sick young one or experience how they grieve over their dead.

I can only say that by their nature, they are limited in that aspect to remain the animals they are. They will become humans if they have as much intelligence as humans. They are not humans. In the Creator›s infinite variety of characteristics, He gave different levels of intelligence to different animals. On one end of the scale, you have the extremes, those with super brains. On the other end, you have humans that have lesser intelligence than gorillas. And so, it is.

We can now understand why the Creator equipped every human being with the human-computer, the brain—intelligence. One of the functions of the brain is serving as a translator or interpreter and as the enabler. A few days old babies looking at the ceiling in their room will not understand what the ceiling is because the interpreter is not mature enough to not talk of being experienced enough to understand what a ceiling is.

Suppose while taking your evening walk, you meet a bully. It is your eyes that did the seeing or collected information about what they saw. The eyes, on their own, can't translate what the data collected meant. Their job is to gather the information and pass the same to the brain for interpreting. On getting the data, the brain will do a quick assessment and determine whether fight or flight was necessary and instruct the body parts involved in the action to take.

The exact process is what happens to all pieces of information coming from the senses. For example, the eyes saw something sitting by itself. From other pieces of data from the eyes, the brain identified the object to be a baby. While processing this information, the ears had reported a sound coming from the object. By now, the brain has concluded the baby was alone crying. Remember, the conscience is not left out here.

The brain has decided for further investigation because the conscience would not allow the individual to stop at that point. Then the brain instructs the parts of the body involved to lift the child for further investigation. While the hands are trying to raise the baby, the nose representing the sense of smell notices some foul odor and reports to the brain. At the same time, the hands reported what they felt, what the brain interpreted to be a wet diaper. These processes take less than a second to execute. The senses by themselves can't assess a situation or solve it. They only collect the pieces of information and pass it on to the interpreter.

A good question here is, "Why did the Almighty God equip every adult human being and, to some extent, other animals with 'working' brains and senses?" You might think He should have given them to a select few who should tell the rest of us what to do.

No! That would be out of character for the God of infinite variety and unlimited knowledge. A child has the same gifts. But we know babies can't make decisions based on their interpretations of the stimuli interfacing with their brains. We know that if a baby witnesses a situation, it can't assess it not to talk of making a decision based on the assessment. These explain why God Almighty gave babies parents or adults to make the decisions for and take care of them.

When a person becomes an adult (including other animals), the person will not need parents or any adult to make decisions for him.

We know that the ever-knowing God did not equip any individual with the knowledge of everything in His infinite characteristics. No one person can effectively be, at the same time, a clerk, a carpenter, a schoolteacher, a lawyer, a cleaner, etc. This fact makes it necessary for us to go to our neighbors or another person for advice or help. To accomplish those assignments, He equipped each animal with mobility.

Yet, the ultimate decision is still left to the individual to make. Considering what He created, it is reasonable to believe that His intention in not allowing an individual to be a master of all knowledge was to foster communal living and make sure we help one another. Nobody or community could claim superiority over others. It allows animals to work together, and nobody isolates himself. And in so doing, humans will work together to take care of His earth.

A person specializing in carpentry should not claim superior to a cleaner or an ironworker, a clerk, etc., because each is playing a part in taking care of the earth. And besides, the person is fulfilling other duties like raising a family.

The man's desire to control and exploit others thwarts the Creator's perfect arrangement. Our teachers, family, religious instructors, etc., trained us on how to think and do almost anything. There is no spontaneity or natural thinking. Thus, our innate creativity is compromised. All these instructors took away the humanness in us. It is like the instructions bandaged our eyes close, plugged our ears and noses, and we are bereft of human feelings.

What is the ‹normal› reaction to shocking televised news like bombings that often killed hundreds of people?

Many will shrug and continue with their busy lives or even continue eating their food as they watch that play out. What is the usual attitude or reaction when some people see someone in distress, or two people engaged in a fight with each other or a beggar near the roadside? What are we taught to do? "Don't get involved in what does not concern you," a misnomer indeed because it should naturally concern every person around when they become aware that another is in distress.

Those behaviors show the level of our learned conformity. Children are taught to keep to themselves and not talk to strangers for fear of foul play. Although initially well intended, this, in effect, deprives them of their humanness and valuable relationship skills. Even elders are deprived of the joy of a child's innocent laughter and play. Sad!

All living things are yoked together in an everlasting bond. Animals depend on the oxygen plants give out. While plants rely on the carbon dioxide animals breathe out. Also, animals (humans in particular) must take care of plants since plants give out all the essential oxygen and food, without which animals will cease to exist. It is clear then why God equipped every person with all the senses, including the brain? The Creator provided these gifts simply because each adult ultimately will be responsible for his own decisions and destiny. Is anybody supposed to be led by the nose by another person? No! We received the same senses and the ability to make decisions and be able to guide ourselves.

This explanation will undoubtedly run into conflict with those who, because of their knowledge, prefer their adherents not to think independently but to soak in whatever they are selling like a sponge in water. It should not be so. Generally, we hear people complain that they did this or that because someone else told them to or they heard it from a source. We must hear and be told things because we live in communities; we live in the world. But we are equipped with the senses and the brains to analyze the information we get and decide based on our analysis.

Also, living things, including plants, are designed to live in communities. Yes, even plants! There are areas where only cashew plants or pear trees, coconut trees, plum trees, oil palm trees, etc., naturally grow. Those are communities. Of course, we are more familiar with communities of peoples, including animals, like the British people, Indians, the Yorubas, the Eskimos, elephants, lions, etc. Even among these large groups, are subgroups?

By now, we have begun to give thought as to what God's intentions were in whatever he created. For example, why a "community" of lemon trees or corn plants or lions, monkeys, or the most common one, humans? You don't see these creations existing individually and scattered all over the earth.

Among plants, think of how natural pollination is; animals think of mating, caring for the young and elderly, and the support and help for one other during hunting or when defending themselves against predators! Amazing! Indeed, the Creator is All-Knowing.

Humans are generally baffled about the creations of God, particularly the ones they don't understand. Therefore, the question, "Why?" will always be there. But now, through reading God's instructions as written in all He created, we understand that there is a reason for the 'why' in anything created, that there is use for everything created. I can confidently say that everything created was to aid humans in completing their task of taking care of the earth. In other words, humans are at the top of the food chain. How so, may you ask?

Any living thing is either an animal or a plant. Humans belong to the animal kingdom. Whatever a plant is doing – providing food, medicine, shelter, etc. – is mainly for human consumption. The same happens with other animals – whether they provide food, services, clothing, etc. – they are also for human consumption.

To provide these may involve giving up their lives. A cow, a plant, may have to give up its life for the good of humans. Is it any wonder that early humans thought we should be appreciative of all life sacrificed on our behalf? Many communities still practice these Thanksgiving traditions today. There is no situation where a human being will give up his life to provide care, food, or services to a plant or other animals. But humans offer care and protection for all living things. They must do so because the very existence of humans depends on the existence of plants and other animals.

To enable humans to execute this all-important function, the Creator bestowed on them the ability to develop the intellect, senses, conscience, etc. These combine to form the Will. It will be explained later why the Creator, God, gave only humans and other animals the ability to compare and make choices.

When a cow, more than six times as big and powerful as a man, is being sold or led to the slaughterhouse, it has no choice. But if a human trader leads another person for sale or slaughter like the Slave Traders did during slavery or Jews to the gas chamber, as done by the Nazis, during World War II, the action would prick some conscience of the onlookers as inhumane. Naturally, the person, the victim, would resist, fighting tooth and nail to free himself.

If a tree is growing near your house, it has no choice but to live there. A polar bear has no choice but to live in the polar region. The same thing goes with a fish that must live in the water. Some people believed that even the 'Angels in Heaven' do not have free Will and conscience. Like ordinary animals, they only do as God bid them. Like the Will, only humans can develop a conscience (the small voice inside) that directs a person to see and do what is right. The conscience imposes the most significant deterrent on human beings.

What might be the reason the Creator gave only humans the ability to develop a conscience, choice abilities, or Will power? The answer lies in examining God's creations and their functions. Anyone who carries out this inquiry will discover that man is the only animal kind that possesses the brains, senses, and Will to rule and take care of the earth, involving setting up rules and laws to guide human and animal behaviors and activities.

Note here that there is no human activity or behavior that is inborn. No! The Creator planted the ability to develop them. These rules and norms that shape human behaviors are external and can be influenced or superseded by a more intriguing influence.

An innocent, loving child raised in an environment where people respect and care for one another will be appalled to see a young person disrespecting an adult or any person.

But suppose this loving person begins to keep company with criminals and continues to participate in criminal activities. In that case, the conscience will change – the person will no longer see anything wrong with destructive behaviors.

The intentions of God are clear—to give the ruler of the whole world – the humans – the ability to discern and choose what is right or wrong. The brain springs to action immediately; materials – data – from the senses are transferred to it. Man can make decisions, choose, or discard.

Only the Creator controls the behaviors of (wild) animals through the Ns factor planted in them. No matter what humans do, a lion will always be a wild beast, operating only according to its instincts. Yes, some external forces can influence natural behavior, but the animal remains what it is. Birds kept in a cage, well cared for to satisfy the owner's fancy; all animals kept and cared for in the zoos; all long for freedom.

It is clear, humans have the authority to rule over all other animals and make decisions about them as they see fit but within some specific guidelines as dictated by the conscience. For example, he can't kill, punish, and abuse other living things just for the fun of it.

How is the validity of these conclusions ascertained? For starters, the Creator designed his systems such that animals depend on plants as plants depend on animals. A man knows that his existence depends, in part, on the existence of plants. Since he is the earth's caretaker, he learns to care for plants.

Consider the emotion stirred up when in the zoo, or even in the wild, someone kills an animal for the fun of it. The outcry over the killing of a gorilla that rescued a toddler that fell over into its cage, and the killing of the beloved lion by the American in Kenya just for pleasure, are a few of many examples. People get punished for willingly destroying plants or forests. People do understand that they have a moral obligation to protect plants and animals.

In many places, people are required to replace any plants or trees they cut down for any reason. There is no question about it that the Creator permits humans or even animals to kill animals or cut down plants and trees when and only when necessary: we need animals' flesh for food and the skins for clothing. And plants and forests also have vital roles to play in the lives of animals.

Some animals are subdued and tamed by God to be obedient to humans. But who controls human behaviors and actions? Not by the thought of the 'inevitable judgment'

after death as legend has it. Or fear of going to 'Hell.' There are no 'Angel' Police Officers sent to us by God to rule and control human behaviors. Besides, natural deterrents like the abilities given to man to make laws and dispense justice do not always work to guide many human behaviors.

The All-Knowing God planted the 'conscience' – the little voice inside human beings. Some people can commit all sorts of evil undisturbed as long as nobody catches them. But dealing with their consciences is a challenge. It is only the conscience that has a firm grip on everyone. Like a person's shadow, it follows a person wherever the person goes.

The conscience being invisible goes a little farther, though. One can lie out of a crooked deal. One can play smart or dumb to get himself off the hook. Even one can employ different tactics to cheat, deceive, or defraud and avoid being caught. Yes, in any of these evils, the perpetrator can scheme and manage to free himself. But how can the person free himself from the ever-nagging 'silent' voice inside the person? The firm grip on individuals by the conscience is what makes guilt tormenting – the judgment by the conscience.

There is no rest or a hiding place for the guilty with the conscience ever reminding and pestering him with his guilt, urging him to ignore his shame and give up his pride and confess his sins. You have likely met or seen a person supposedly very rich or famous and happy. And suddenly, he committed suicide or disappeared from the radar. Some could go as far as undergoing plastic surgery to disguise themselves.

They go through those actions in an attempt to hide from the law and themselves. He could not face himself; they can run but can't hide from their consciences. Many wealthy people amassed incredible wealth by all 'means' possible. But at a certain point in their lives, generally, towards the end of their existence here on earth, they decide to give all that they accumulated away. Why? It could be due to the goodness of their hearts. But more than likely, it was the work of 'the silent judge and jury' fused – the conscience.

It is the conscience that creates the human kindness we encounter. The conscience is the little voice inside that would always caution against doing anything wrong but encourages godly actions. This type of action does not seek reward; neither does it need publicity. Why? It is from the Creator.

I remember what happened when I was given the government scholarship to go to America and further my education. But there was a make-and-break condition – the signing of my 'affidavit'. Initially, I thought that was the easiest part because I had people well situated to get the job done.

To my greatest surprise, these people failed me. But I had an in-law, whom I never dreamed would appear in the equation. When he saw me weeping because I was sure to lose the scholarship if the Affidavit was not signed as required by the Authorities, his conscience went to work questioning how he could stand by and watch me lose the 'once-in-a-lifetime' opportunity. He had only one choice!

I still remember a time in the US when I could not find employment or instead got hired, my second degree in Civil Engineering notwithstanding. You might be wondering why? I did not possess the legal papers.

After a series of disappointments and losing a lot of money, all stemming from my legal status, I became bold. Giving up was not an option. I decided to be upfront with my situation in the following interview and choose my words carefully to appeal to the 'Human Kindness' – the conscience. Surprisingly, it worked. I could not believe myself. I will ever remain grateful to him.

What would be the stat f our earth if humans don't possess the conscience? Of course, the Creator knew everything! He knew the disaster that would occur.

Making choices is a different story. To choose, one must be able to discriminate between or among the available options. If the choice is between two or among many delicacies or beverages, or which clothes to wear for a party, or which beautiful woman, a handsome man, to marry, etc., there must be a list of criteria guiding one to the choice or choices. Luckily, the Creator provided the potential options in abundance. In other words, there is no excuse for not finding whatever is your choice.

There will be an inevitable rejection in this deal. Rejection or acceptance will, many times, follow a choice. One of the two is always the case. Humans possess Free Will to be able to choose or reject. Should this cause dissension or discontentment?

Should this be good or bad? The answer will depend on the intention, understanding, and or purpose of the choice or rejection.

The dance committee will choose only one woman to be the 'Dance Queen' at a dance party. Also, in a presidential election, a country will choose only one person as the country's president. And for an air-plain pilot, a selecting committee will choose one person among several to be the captain. Should there be a wholehearted acceptance of the verdict? The answer is as complicated as the question itself.

Suppose there is a contest to select one young, beautiful, and average woman to fill a position. You have, say, ten women that fit that description. Then the selecting committee will continue to raise the bar until a winner emerges. Raising the bar until they find the best choice is how people of conscience, have the 'fear' of and respect for the Creator at heart.

It does not happen that way very often in real life. In normal circumstances, in an impasse, as described above, the human factors must be injected into the decision-making, ignoring the 'Silent Judge.' One of the judges might have been taken out to a restaurant by one of the contestants; one of the contestants could have a mutual friend with one of the judges. Such outside influences unrelated to the set criteria would cloud judgments. The people involved will question the verdict or choice, leading to discontentment.

In my young days, I used to have a friend. That person was fashion-conscious and owned a lot of clothes. On Sundays, we attended service together; generally, the first one at eight am. On one particular Sunday, we had guests from different localities. Being fashion-conscious, my friend started getting ready at 5 am. She needed more time to try on different clothes to choose one that fitted best. At that time, I was still asleep. When 7:45 am came, my friend was not ready, and I had to leave for fear of going late. My friend was late for the second service at 11:00 am. My friend had a problem with making a choice!

From those scenarios, it is easy to see that making some choices could be easy and harmless, and others could be tough and hurtful to the rejected and could lead to unacceptable behaviors and consequences. Why should there be a problem with choosing and rejection? After all, between two prospective candidates, only one will be the choice. The first scenario identifies the human factor – partiality. That outcome was not of the Creator.

Over the years of human existence here on earth, people encounter making choices, rejections, and outright discrimination. These behaviors have caused more harm than good. It is a blessing that the Creator allows His creatures to choose from the abundance of His gifts. One can choose from different types of oranges, mangos, apples, vegetables, peppers, tomatoes, etc. (see figs. 7&8). Making choices have some elements of discrimination built into it. There is no other way to choose. Then, why are some aspects of discrimination bad?

All types of discrimination are suitable except ones made unfairly, such as bigotry and hatred. Rejection is a mild form of hate:

When people reject something, that means they would prefer something else. One might prefer Cole slaw to a salad, like one might prefer Okoro soup to Egusi soup. In these cases, it might be a matter of taste or the ingredients used. The aversion is much higher when you hate something, i.e., you would prefer not to see the object of your hatred exist on earth (or at least in your presence).

During World War II, the Nazi Government in Germany hated the Jews to the point they decided to exterminate or remove them from the earth's surface. The same situation existed during the slave trade: a whole cargo load of black people could be emptied into the ocean to avoid detection or capture by British Slave Monitors. No sane person can even treat dogs, goats, cows, etc., that way. Did that not show the level of hatred?

Many white/black peoples dislike many black/white peoples. The choice is likely to be a matter of preference, taste – behavior, attitude. Also, some white people prefer black people. Again, few white/black peoples hate some black/white peoples. One can see the evidence of this in the activities of racist organizations like the KKK.

How would anybody prefer or wish an earthly object not to exist on earth? How would the creator and owner of that object feel? Here, they disobeyed the creator, which makes hatred among the worst behaviors and should be avoided at all costs. The Divine consequence of this behavior has led to much of the strife and unhappiness here on earth.

Discrimination is inbuilt in all animals for good reasons, but the extension – hatred, is acquired with time. I have described discrimination in more detail elsewhere in this book. It is inbuilt in all living things; then, it must be of the creator! The Creator is All-knowing and can't make a mistake. It is better to cross-check the reasons proffered in this book as to why the Maker of all things, planted in humans, the ability to develop discrimination - go back to nature.

Since humans can't go to the creator for questions and answers, going back to or studying nature in our orbit is the only way to obtain clarifications about any creation we have problems understanding. A friend was having a wedding party. He checked with the meteorology department the chances of it raining on the party day to assure the success of the wedding. He received a 60/40 percent chance it would not rain. My friend looked at the cloudless sky and convinced himself it would not rain. Low and behold, halfway into the feasting, it rained and ruined everything. He complained of why God did not see the event and stop the rain. Did the complaint make sense? Of course, no!

When the Maker created humans, He planted (the ability to develop) the brain and its senses. He did not leave them without resources and guidance to help them navigate their ways here on earth. He knew that there would be what earthlings call difficulties. My friend knew that if the chance for rain was likely, he should have made the necessary arrangements to avoid the disaster.

No! God would not allow his ceremony ruined! Perhaps he was born-again or prayed five times a day. Everything needed to avoid his celebration destroyed was already provided by the All-Knowing. Is rain a bad thing? Who does not know the need for and use of rain? At that time, as in other times, many living things needed the rain. Of course, God does not play partiality and is aware of all situations – He had already provided the intellect and senses. The grieving man needed to talk to himself, not to blame God.

Another good example is the UVA from our sun's rays, which will cause skin cancer at high doses. But peoples in the equatorial regions of the world are always soaked in the UVA. It would be reasonable for them to complain to God to remove the Sunlight to avoid skin cancer. Of course, there is no need for that—they rarely get skin cancer!

The Maker of all things had a good reason to create the sun. Without the sun, there will be no life on earth. To counteract the effects of the UVA, He put Melanin in the skin of those, He assigned to the equatorial regions of the world where there is too much sunshine. Melanin prevents cancer of the skin. Nothing is left to chance. God is God, the Ever Knowing!

The problem we experience here on earth emanates from a lack of understanding of and trust in the creator's knowledge and power. Otherwise, we should simply follow the dictates of nature. God knows all the desires of our hearts and provides them as He sees fit. Even in the circumstances, He allows us to control; He provided the resources and means (the brain, senses, consciences, etc.) to help us in whatever choices we make.

We certainly got the knowledge of the Creator wrong right from the beginning. Or else, the earth would be a place of enjoyment as the Designer intended. In any situation, the creator is in control, and for any creation, the creator has a good reason. All humans need is 'time out' to go back to nature and read His intentions written on the creations and adhere to them.

Does He expect hours or days of prayer to remind, persuade, coax, or force Him to do anything? There is no room in His creation for intercessors, prayer warriors, going to another earthling to confess and receive forgiveness for sins committed. Since we know He wants us to love one another or be our brothers' keeper, does it not make sense to believe that He intended that one goes to the person wronged for amendment and seek forgiveness? We got it wrong at the beginning, or our religious teachers misled us.

At this point, we turn our attention to the whisperer, the opposite of conscience. The whisperer is the 'evil' small voice in every person. This little voice can't see anything good in anything, not even in its host. It is also a product of the environment. Until one silences this evil voice, one cannot be truly free.

You invited a friend to dinner. Twenty minutes after the appointed time, he had not shown up. What did the evil 'small' voice say? "Maybe he forgot. I knew from the beginning he showed more interest in my girlfriend than me. I don't think he liked me" – all negative.

The man was having a car problem while on his way to meet that appointment. Being in the middle of nowhere with no phone to make a call, he had to walk more than two miles to get help from a gas station. Think about that! Have you ever distrusted a person for no reason or met somebody you had never seen before but for no apparent reason hated or disliked the person? Yes, this is the work of the silent whisperer, created in mind by past experiences in our environments.

One day a man with his three children was waiting for his flight to return to his home. The man was sitting by himself with his hands folded. The children, left to themselves, were disturbing and pricking the nerves of other passengers. Many of the passengers were sitting there, enduring the cacophony. And they were guessing why the father was incapable of controlling his children. Some went as far as to think whether the children were his in the first place. How could the children be so out of control even in the presence of their father?

One of the passengers, fed up, took the initiative to confront the father to inquire why he was letting his children disturb their peace? The grieving father took the opportunity to inform the disgusted passengers that they were returning from the burial ceremony of the children's mother, his wife. And for the whole of the previous week, they all were packed in one room to endure their loss. "You can see why they needed the breathing space," he concluded. On hearing this story, some of the passengers took out napkins and wiped their eyes. Some passengers reached out to the children and consoled them. One can observe the conscience, and it's opposite the whisperer, at work here.

Why must we condemn first before even hearing the story? Why do we first look for what is different between 'me' and 'you' or "why? I'm better than . . ." the other person?

As soon as one encounters a stranger, there is a rush of curiosity to know who this person is. The level of inquiry will depend on perspective, mood, and circumstance. Animals, particularly humans, react that way in such situations. A relationship—acceptance or discrimination—will emerge depending on many variables, the highest on the list being skin color and appearance. Sexual orientation is further down the line.

Have you ever observed the feeling of a 'big' man well-groomed in an airplane flight, sitting next to a person he feels was beneath him? Just watch his attitude towards the 'thing' that calls himself a human being and notice his air of superiority.

Have you observed the attitude of some flight attendants serving some passengers who did not look like them? Check out the attitudes of the functionaries of banks, market places, courts, schools, officers in the streets and freeways, and even the clergy in many areas of 'worship.' Have you not observed the treatment some people receive because they were from a different ethnic groups, have a different accent or speak a foreign language or dress differently?

There is no way we were taught right about who the creator is or what He expects of us! No! He is the creator of all things; without Him, there will be no life on earth. He created our beautiful planet for our enjoyment. The sunshine, rain, animals, and plants, He created for our use and enjoyment. He only desires us to take care of the earth, including loving all He created not for His benefit but ours.

It makes all the senses that He established the commandment, "Love your neighbor and take care of my earth as I have shown you." That law was not only for black people or white people, or red people. No! It is for all of us. Any time humans disobey the creator's laws, there is a consequence – the Day of Judgement. We know we consistently violate His laws. Yet, we wonder why the unhappiness here on earth. These are the consequences of not obeying God's commandments: mistreating each other or not taking care of the earth.

On the other hand, some people spend all their lives fighting for the 'little' guy. Any injustice people perpetrate pricks their conscience. In most cases, they don't ask for pay, recognition, or praise. They instinctively know that the purpose of life or our duties here on earth is not to amass wealth, ignore others, and care for self only.

It will be better for those that practice selfishness to visit hospital emergency rooms and observe those with horrible injuries or at the point of death. It might be better for them to visit hospital sick rooms, where the very ill lies, praying for death. Also, one can attend a burial ceremony, particularly that of the rich. These people have reached a point where wealth has no value – it could not save them.

At this point, they even have no memory of their cars, mansions, money, or any of their earthly possessions. At this point, they realize (unfortunately too late) that God is God. Remember, previously, and because of their wealth, they had been tempted even to see themselves as God. Nobody could touch them. But when they find themselves at the end of their ropes doesn't the whisperer in them ask the question, "Why does God allow bad things to happen to 'good' people." Yes, 'good' people they call themselves!

Despite all the good things God provided for his creatures with nothing left out and how He takes care of His creations, it is surprising that some humans still hate as much as they do. Some do not carry out their hatred physically. The unfortunate part of this behavior is: as Martin Luther King Jr. said and I paraphrase, if any person stood by while evil was perpetrated and said nothing or did nothing, the person is equally culpable in the crime as the person who committed the crime. The lesson of this saying might be why the Dali Lama cautioned, and I paraphrase, if you don't want to help someone, at least don't hurt him.

In my analysis of the family set-up, described later in this book, told how hatred entered our otherwise beautiful earth—a learned behavior, the teaching of which started in the family, exacerbated by socialization outside the family and past experiences? Can anything be done to correct this anomaly? The answer is, "Yes."

First, humans are generally good and so created. And we know that God's creation is sound and complete. And He wants His creatures to live here forever and happy. We know these things to be true because of His provisions, care, including all the beautiful things He created, particularly His impartiality.

Humans must understand and accept the infinite varieties in His creations—multiple shades in color, different levels of understanding, height, weight, temperament, etc. These creations are of God and, therefore, good.

We must accordingly use the brains and the will He gave us to make choices and determine our destinies. We should look around us to see the extension of the creator in everything. Then remember that everything belongs to Him and that He wants us to share them among us equitably. Luckily, He provided us with a series of deterrents/reminders and how to deal with them.

At this point, let's examine love and hatred. There is a thin line between 'love' and 'hate.' The line is very thin, indeed! Hate is the direct opposite of love. But it does not mean that where there is no love, there is hate. But love is everything. There are two types of love – 'Natural or Selfless love.' And the other is 'materialistic or 'selfish love.' Is natural love the same as infatuation, or is it different? Is selfish love the same as pretentious love? Though immaterial, these different types of love can easily be differentiated or described. As likeness or fondness grows with time, it becomes love.

Love, therefore, is a deep feeling of fondness that turned into devotion. When you love, you can't help or control yourself. It can make one behave irrationally. There must be an element of time to develop love; it can't be instantaneous. Young peoples and those who have not learned this do not factor in the time and confuse love with infatuation.

If we surround ourselves with genuine love, our earth will be a happy place as the creator intended. We can only be happy by performing our Creator-assigned duties, which we should know to be and call prayer. God made carrying out these tasks easy by providing all one needs to do them and the reward of fulfillment in their execution. I will now discuss "guilt," the silent killer.

If one is proved to have done wrong, then he is declared guilty, followed by some form of consequence, e degree of which is determined by the enormity of the offense.

One can unjustly inflict pain using means ranging from verbal to physical. Consequences of guilt can range from a simple warning like rolling eyes to prison terms or even a death sentence. Remember the role that the deterrent – conscience— plays in all of these. It forces one to confess, accept the verdict, and serve a sentence. Doing this is the only way to go.

Suppose authorities did not catch the sinner? Yes, he will do all sorts of things to cover up the sin. No! That can't work. It cannot be acceptable and transparent. "Why?" you may ask. The ever-present conscience is there and at work, keeping a constant reminding schedule. The never-ending reminding of the conscience will ultimately induce one to hide oneself from the victim or the law.

The deterrent called conscience is about the only thing that can keep humans' behavior in check. Sometimes, you can see a grown man shedding tears, confessing his misdeeds.

Consider people who surrender all they owned for public use. It is better for them when that happens. It brings relief and contentment to the soul, joy, and happiness to the world.

In some instances, the guilty will refuse to yield to the nagging of the conscience. Then the situation will become severer and dangerous because the conscience will instead double its efforts. The person will eventually surrender to the pressure. The 'day of judgment' started right from the onset of the crime. And the consequence is proportional to the enormity of the offense. There is no other way to get out of the evil and its results than making amendments to no other person than the victim. Why? Because the divine law is to take care of living things as you take care of yourself.

These explanations notwithstanding, many people still wonder why the unhappiness on this earth otherwise created to be a Paradise. The creator provided everything to make that possible.

Please revisit the figs. 7&8. The images are used to explain the 'infinity' and 'variety' characteristics in the creation. Considering the 'love' and care, He has for His creatures, the thoroughness in His creations, His inability to make mistakes, or forget anything, it is reasonable to ask if the creator had other reasons for the infinite variety.

I've always wondered how 'mothers' got the idea of variety when preparing food for their families, particularly families with picky hard-headed members. It must have come from the Great Teacher, the creator. He knew the type of animals He created – some, discriminatory, and hard to please. Therefore, the apples, oranges, vegetables, ears of corn, etc., were created in different colors, sizes, etc., to make sure His children, no matter how picky, would find what to eat and enjoy. To be sure of this, all the foods carry all the nutrients as He intended. So, choose any vegetables, fruits, etc., many of the same nutrients are provided in them by the creator. (See figs. 5, 7, and 8). Should there be any doubts that the Creator is God and loves His creations!

Chapter 7

FOOD

God created all things for the enjoyment of humans and to aid them to complete their duties here on earth: preserve life, extend their kinds, and take care of the earth. Nothing more and nothing less! Whenever we step outside these functions, there are problems and consequences. I'll describe later how He built these three functions into His creations.

At this point, a review of the points made earlier in this book is in order. We know that the earth, among all creations, is the only place humans inhabit or life as we know it exists. There is no other known place. We are sure of this because the earth is the only place equipped with all those living things' needs. The importance of these is evident - required to aid them to fulfill their functions here on earth. The supply of these needs is so programmed, they are self-perpetuating and will never run out.

We now understand that the only material required to fuel the human 'machine' is food, not any food but the type created by the Maker. Since humans and other animals must eat, the Creator placed the food everywhere on their paths and planted the urge to eat. As they wandered around, prompted by the Ns factor, they started eating. We also know that they settled and began to produce the food themselves in their localities.

Think about it: an orange tree, mango tree, pear tree, etc. will continue to yield fruits year after year or season after season until the end of its life, and before that point in time, their seeds had grown to maturity and begun to repeat or continue the functions of the 'parents' (see fig. 6). The most amazing part is that these trees, fruits, seeds are not one type but uncountable, as usual. Think of the different kinds of vegetables, tubers, etc.?

They are so abundant we hardly notice they are there or ask about who put them there and for what purpose. We need the food to keep us alive and then be able to fulfill our duties here on earth.

You don't need any person to get your food. The requirement is you must 'go' and get it yourself. They are free and available.

The only way to derive value from food is by eating it. You can't derive food value from peanuts, for example, by looking at it or feeling it. As you are required to get the food, so you are required to eat them. The Maker knew beforehand that humans would need a system to derive benefits from the food they eat. The primary mechanism He provided was the digestive system. All animals were provided with this system, with modification in some animals because of the food the animals eat.

Many animals chew some types of their food before swallowing. You will be surprised by what you see on a second look at your teeth provided by the creator for cutting, tearing, and grinding the food. You will marvel at the design of the surface of the premolars and the molars to suit the type of work they do. Some animals swallow their foods directly, while yet others drink or suck theirs, etc. Humans and some other animals do both at various times in their development.

Plants have a different system because of the type of food and how they 'eat.' Plants don't need the digestive system: the roots gather nutrients and transport them to the leaves via the stems. The leaves, using a process known as photosynthesis, manufactures the plant food. Then another system transports the food directly to the cells.

Except for humans, all the food these living organisms need throughout their lives the Creator has already provided. All they need to do is simple: 'Go and get it.' Ants, birds, goats, fishes, and many other animals are not equipped to plant. The inability of these animals to provide their food places on humans the task of producing and preserving the food for all living things. Only humans are equipped to plant and sow. The seeds and fertile soil are provided free of charge.

Consider the following: To be created, live here on earth; eat the food provided to live; must be done by all living things. But they have no say in the decision. It is reasonable to believe these are divine commandments. The miseries we live with here on earth are the consequences of failure to comply with the Creator's laws.

The food sustains life, but how? The nutrients in each food group are for specific functions of aiding living things to perform their duties here on earth. Hence, to look for or produce food, protect the self, and the members of the family, animals, will need strength.

Which food nutrient solely takes care of that? Carbohydrates, as available in fruits, seeds, crops but mainly in tubers.

What about fighting diseases or nourishing and growing the body parts? The Ever- Knowing Designer has created fruits, vegetables, oils, plants, animal proteins, etc., to take care of those. Remember, the supply of these is forever and free. But all living things must eat the food as the Creator decreed to set the nutrients free for absorption.

It is only humans that are equipped to plant, cultivate, and reap. Why? The ever-knowing Designer knows that human food (including those of animals and plants) would run out if nobody replenishes them. The humans were instructed (by the Ns factor) how to keep their food supply available if they were to survive on this earth, i.e., forever. Remember that humans possess the brains – the seat of knowledge. Other animals have the same things given to humans except for intelligence. This difference is big because humans know to plant and preserve food and do many things that other living things simply cannot.

The Almighty Designer knew that our living on this earth would be a journey - we will get tired, sick, and hungry. We will need companionship, love, protection, enjoyment, etc. The Creator provided all the means to satisfy these needs. For example, human and animal 'babies' can't protect themselves, provide their food, or teach themselves the ropes. The Maker knew and took care of those by providing them with 'parents' waiting on them right from conception. In the womb, the umbilical cord attaches the baby to the mother for its food. In plants, the design is different: with the seeds ('the young ones'), the Creator attached their food on them through the umbilical cord to the 'mother.'

After humans and animals eat the juicy pulp of an orange fruit or the fleshy part of a mango, the seeds are generally thrown away or discarded. We know that the Creator's intention is for that seed to take over from the 'mother' someday. The seed can't find or provide food for itself. Neither can the 'mother' do that for it? So, the Mighty God attached the food to it – all the food the seed would need before finding food on its own.

The air and sunshine they need are already available. Seeds have their water already, and when they need more for growth and functioning, the Maker has already made arrangements for that. The adults of these creations have their types of brains guided by the Ns factor.

Gas, oxygen, and carbon dioxide are other essential elements without which nothing will live on this earth and be able to fulfill the instruction: "live on this earth." These two gases exist in nature and are byproducts of photosynthesis or respiration. Animals inhale oxygen and use it for different functions in the body, e.g., oxidation of the blood. The byproduct of this process is carbon dioxide. Plants use carbon dioxide in making their food in a process called photosynthesis. The byproduct of this process is oxygen. These two gases are vital in the lives of all living things.

God, the Maker, provided to animals the respiratory system to derive benefits from the gases. The Maker also provided other methods in all animals, though modified to suit. And there is no option. Again, the word "free" comes in.

We use the word 'free' loosely because we don't appreciate it in its entirety, particularly as it relates to the Creator. If we had an iota of understanding of the meaning of that word, we would then begin to appreciate the mightiness of God, the care and love He has for His creatures. Then our knees will bend in adoration, and we will obey His commands to love all His creations and take care of the Earth as He did to all His living things.

Unfortunately, we got that concept wrong right from the beginning.

Let's put the gas, oxygen in context to understand how the word "free" applies to the Creator's gifts to all living things. In all hospitals, oxygen is administered to patients for different reasons but particularly during surgery. In such a situation, oxygen can run up to $10.00 per hour for adults and $6.00 per hour for children. So, for a whole day of twenty-four hours, the bill will be $240.00 for an adult. If we follow this pattern of calculation, the amount for one month becomes $7200.00. The bill will become $86,400.00 for a one-year supply. The amount is for only one person for only oxygen gas, not including other gases living things used in surgery or otherwise. Then how much per year will different types of foods, fruits, seeds, vegetables, meats, water, oils, milk, air, sunshine, etc., cost? All of these are provided by the Maker freely! Amazing! Those who understand are right to believe that the Creator God is kind and generous.

Water is another element, without which nothing will live on this earth. A good proportion of an animal's or plant's bodies is water. Since this element is essential to sustain life, the Maker provided it in abundance. It is not for sale but, again, free.

Most of the food living things eat contains water. There is always precipitation. There are streams, rivers, seas, and oceans. Some of these water sources may not supply fresh water suitable for human and animal consumption, like rivers, seas, and oceans. In this for instance, God's Water Cycle system purifies the water for all living things. The process makes the amount of water created remain constant. As we shall read later in this book, the Creator keeps the earth going because of the Creator's Recycling system. When any plants or animals die, as noted before, they are used to feed the living. The phenomenon is fantastic! There is no waste in nature.

While the Creator's provisions are in abundance, yet He forbids wastes. We know this to be true because of the disastrous consequences that occur when living things waste resources. When a piece of land is overgrazed or over-farmed, its natural nutrients are depleted. We don't dig up the soil and 'throw' it away. Nature has in place a way to rejuvenate and replenish the soils.

Unfortunately, humans follow their ways, not nature's, to solve the problem 'How' do we solve this problem? Generally, we use chemicals. The producers of these products allege that we will run out of our fertile lands, beef, milk, and even chickens and egg supplies without growth hormones and other chemicals. What is the truth?

Foods humans produce using chemicals our body sees as foreign materials, unsuitable for use by it. They, instead, have caused inflammations, various adverse reactions, and diseases. We know that the motive for using chemicals by the producers of these not-too-good foods is greed and wastes. The Creator's food supply will last forever if we can follow His rules! The only way we show our love for God will be to protect and take care of all His creations. Greed will disappear, also waste if we use what is necessary, just what we need, like other animals.

Chapter 8

THE FUNCTIONS OF THE FOOD
THE CREATOR PROVIDED

Let us examine the functions of the foods the Creator provided. By so doing, we may understand why He commanded that all living things must eat them. The Creator provided five food groups: protein – from seeds, meat, and fish; carbohydrates – from grains and starchy foods like rice, potatoes, yams, cassava, oats; fruits and vegetables; fats and oil; and milk and dairy products.

There are different types in each group. The same food groups are available to all living things on earth. In His infinite variety of characteristics, the food items in each group may differ from country to country. Therefore, the types of food items that contain carbohydrates in Nigeria may be different from foods that contain carbohydrates in another part of the world. In Nigeria, for example, yams, cassava, and coco-yams are the food items that supply mainly carbohydrates. The food items grow in equatorial regions of the world but not in temperate areas. Therefore, you can't see farmers in Britain or any other country in the temperate regions planting the same food items that have their homes in the world's tropical areas. So, somebody from or going to Japan bought the African yams you see in Japan from Nigeria or any equatorial regions. Let us digress a little.

Just picture the shock and wonder on the faces of the ancient peoples when they saw their first seed germinate and watched it grow and produce the same kinds of fruits they ate previously. They must have prostrated themselves in gratitude to a loving Provider. They saw in rivers, lakes, etc., an abundant supply of fishes and other sea animals to add to their wonder. They did not need a fishing trawler. With bare hands, they could catch enough fish to satisfy their needs.

Now, they could gather fruits and seeds, hunt wild animals, or, in their localities, plant and sow. Food was everywhere. In time humans started battering for their food, and later still, they started buying their food. These exchanges are proper and necessary because no one person or family can produce all their food needs. The Creator intended to make sure we are united in taking care of one another and, by extension, taking care of the earth.

Let me examine in more detail why the Creator made it mandatory that all living things must eat? As usual and the only way to explain this is to explore the food He created. As stated earlier, there are five food groups: Carbohydrates, Protein, Milk and Dairy Products, Fruits and vegetables, fats, and oils.

Carbohydrates occur in starchy foods like rice, potatoes, yams, oats, etc., and may appear in smaller quantities in other food types. On closer examination, you will see that carbohydrates are in more foods than we think. Bananas, pawpaw (papaya), carrot, and many vegetables contain carbohydrates. Note: Some of these carbohydrate-containing foods may not be the same in every region of the world or may not be the same among all ethnic groups of the world. The vital thing to note here is that the Mighty Creator distributed all the food groups to all the peoples of the human family and animals. Amazing!

But why is this so? Why must all animals consume carbohydrates? We know carbohydrates supply energy, and energy is required to do work. The connection is clear: All living things must "eat" and must "go and get the food" – physical movement, action. This commandment is for all living things. The only exceptions are those unable to fend for themselves, like the 'seeds' and 'babies.' He attached the food on them.

The Creator designed all living things and, in the case of animals, provided the means to fulfill the order, "go and get it,"; "interact with other creatures of your kind," and in the case of humans, "take care of My earth." We know that not every type of food can supply energy-giving carbohydrates. But those that do are in abundant supply and free. He knew all our activities require the use of energy. While sleeping, playing, eating, etc., they need energy. It is no wonder the energy-giving carbohydrate He put in almost all the food types.

Another food type is protein. What does protein do? This nutrient is essential in the body of all plants and animals. And why? It is mainly responsible for building and repairing tissues and aids the immune systems. In the body, it is called amino acids. God put it in our everyday foods to be readily available, like nuts from plants, animal meats, milk, etc.

This food group, protein, is undoubtedly essential for the function of the body. The Lord God knew that there would be wear and tear in the process of performing the duties assigned to us. He is the All-knowing spirit, and His agents (angels) are also spirits. It would be odd or maybe frightening to send His spirit agents to come to us to build or repair worn-out tissues or body parts as car mechanics do to broken-down cars. Unlike human-made mechanisms, His 'machines' must be self-repairing. Hence protein became necessary and was created abundantly in our foods to do the job.

Another food group is Fruits and Vegetables. These two generally go together because they supply almost identical food nutrients called vitamins. "What can the body do with or without the Vitamins?" Vitamins are food nutrients the body uses to function and fight off diseases. They are two groups – fat-soluble and water-soluble – and thirteen types in number, like vitamin A, B's, etc.

Why did the Creator make vitamins? Scientists have proved that diseases will disappear if the body obtains all the groups in the right proportions and their natural forms. And this is the main reason why the Designer created them in the first place. Many expensive prescriptions are mostly concentrated vitamins.

Because of their importance, the Lord, God put them in our ordinary everyday foods available to all His animals: Foods like orange-colored fruits and dark green leafy vegetables, our free sunshine for vitamin D, milk, meat and dairy products, whole grains, nuts, and seeds, all supply various form of Vitamins. Of course, these are readily available to all animals. There is no time when there are no fruits or vegetables in season. Because the Creator God knows their importance to His creatures.

Since I am not in the science profession, I can't advise on the chemistry of these nutrients. This writing aims only to examine God's creations as seen in our environments and try to understand the reason behind their creation; detailed inquiry, in this case, is unnecessary. I will briefly examine two types of vitamins – C and D.

When red blood cells combine with vitamin C, that will create a disease-fighting machine. This machine can cure or eliminate many diseases like cancer, heart diseases, etc., because it is an antioxidant that can prevent cell damage by providing oxygen to oxygen-deficient tissues. The Creator provided vitamin C to all organisms and animals, apart from humans. But in His infinite wisdom, He provided vitamin C in many common foods; humans eat, as pointed out above. How will animals, including humans, survive without vitamins?

Just imagine what humans and animals encounter in their everyday living – germs, diseases, different types of ailments, the common tear and wear, etc.

Vitamin D is different from other types of vitamins because you cannot get enough of it from food alone: This vitamin is abundant from sunshine. The body makes it when it absorbs free ultra-violet radiation from the sun and, in the presence of cholesterol, turns it into vitamin D. Scientists have shown that Vitamin D, in enough quantities, can eliminate about 50% of skin cancer cells. This claim is valid because skin cancer is rare among peoples living near the equator, where the skin constantly soaks in the ultraviolet rays. Also, one of the causes of depression is Vitamin D deficiency.

I will discuss the case of sunscreen and skin cancer. Many people living in North America, Europe, and in many temperate regions of the world have light skin tones and generally use sunscreen to shield (they believe) their skins from "harmful" (cancer-causing) effects of the UVA rays. Nothing can be far from the truth. Sunscreen blocks the absorption of the beneficial UV rays and, by so doing, reduce the production of Vitamin D considerably.

The Creator, God in His infinite wisdom, knew that peoples He assigned to live around the equatorial regions of the earth will absorb too much of the UV rays, which will cause skin cancer. But these dark-skin people do not contract skin cancer, despite too much sunshine. Why? Because, as stated earlier, the Creator provided their skins with Melanin, a natural sunscreen that blocks most of the UV rays passing into their skins. The little that is allowed in is enough to produce Vitamin D since their skins constantly soak UV Rays. This substance, Melanin, is also responsible for their dark skin color. (***Please google all the information stated above***).

People believe in some quarters that the adherents of the Mormon Faith used to say that the dark skin was a punishment from God because of the sins colored peoples of the world committed. A curious person might ask what the sin was (is). The belief cannot be accurate, and the saying, at best, pure ignorance. They didn't know that the Creator is the God of infinite variety. He made His creations of different colors. Look around you. Nothing is one. Only the Creator is One!

God does not create one type of anything. Only He is one! Look around you! All human beings of the world are not of the same height, say everybody to be six or four feet tall; all the fingers the same length; all the flowers the same smell or the same color. Why should any person expect the skin tone of all humans to be of the same color? Scientists also determined that Vitamin D is one of the disease-fighting machines, particularly against Tuberculosis.

It is clear why the Designer created fruits and vegetables that contain vitamins, the disease-fighting machines. In their journey to accomplish their Creator-assigned duties, animals encounter different hazards that can interfere with their ability to complete their assignments. The Creat r knew and provided them with remedies. He packed these vitamins in the foods He created for all living things. One example will suffice to show how the Creator compensated for the harmful effects of the sun's rays.

The sun is involved in many things that help all living things sustain themselves here on earth. It is engaged in recycling water, which at the end of the day results in precipitation. Remember, the water created at the beginning of time is the same water and quantity we use even to date. Amazing! The sun rejuvenates the earth with its warmth, ripens fruits, and helps to provide food for plants and, by extension, animals.

The body will make the all-important Vitamin D from the sun's Ultra-Violet Rays (UV); the body will make the all-important Vitamin D., as discussed earlier. But in a high concentration, the sun's UV can cause skin cancer. The Creator, in

His infinite wisdom and 'love' for His creatures, provided the Melanin to the skin proportionately with natural human habitat to counteract the effects of the UV.

I want to suggest here that dark-skinned people who, for different reasons, find themselves living in temperate regions instead of equatorial regions should remember to expose their bodies to the sun's rays whenever the opportunity presents itself. Why? Abundant sunshine is designed for their skins. But, in the temperate areas, the body is always covered, mainly during the long periods of cold seasons with little opportunity to expose the body to abundant sunshine. On the other hand, fair-skinned people, particularly when in equatorial and desert regions, must limit their exposure to the sun's rays because their skins are not designed for abundant sunshine.

The last food group is Fats and Oils. Fats and Oils are essential to the body for various reasons. Our bodies use fats and oils to build hormones, keep our skin healthy, synthesize fat-soluble vitamins, provide energy to the body, and are the brain's primary energy source. Foods rich in fats and oils are whole eggs; nuts (like the almond, pecan, walnuts, cashew, peas); seeds like the sunflower and flaxseed; Avocado; Olive oils; red palm oil; fatty fish like the mackerel, tuna, salmon; and grass-fed butter.

God, being God of variety, did not provide the same types of food in all the regions of His earth. But, since the bodies of the animals, He created need fats and oils to function,

He provided the type of foods that can supply fats and oils in every region of the earth. He is the ever-knowing Go He does not forget.

My wife and I spent two years in Ghana, West Africa. We had the opportunity to spend some time in the Northern part of the country. The cooking oil people used in that region surprised us.

In Nigeria, people use palm oil, groundnut oil, and varieties of vegetable oils in cooking. Also, in the United States, where we live, except for palm oils that few communities use, other people use various varieties. You can imagine our surprise when we learned that Shea Butter oil was also a cooking oil and that Europeans had been using shea butter in chocolate for many years.

There should be no surprise about that, in part, because Palm trees that produce palm oil did not grow there. But the Creator God did not forget that the human machine He placed in that region would need oil to function. Hence, He created the Shea Butter oil to be used in their locality and traded by exporters who realized its nutritional value. It is clear why food was created: to provide sustenance and ward off diseases so that all living things would be able to survive on earth happily and forever with the ability to complete their assignments here on earth.

As stated before, the Creator knew the hazards His creatures would face. So, He provided remedies and medicines to deal with those hazards. He gave the intellect to the earth's caretakers. It is no surprise that caretakers followed the example of their ' parent,' the Creator, to find and use medicines to fight ailments and their causes. To achieve this, humans have no other choice but to use earth materials. The earth caretakers cannot create.

Only God can create. There is a problem or retribution when humans try to 'create' by changing or altering the form or composition of any earth material. We know that when humans mix two substances chemically, they will always form some synthetic (unnatural) compound. If humans administer this compound into any living thing for whatever reason, it will be rejected (or have side effects) by the body in some way (even if only slightly). It is not natural. So, these outcomes warn humans to use all earthly things in their natural forms. We are aware of the disastrous consequences of using any earth material in its raw form or altering the natural composition. We can safely agree that the unnatural use of Earth materials is the second source of unhappiness here on earth.

At this juncture, the role played by plants and animals as far as the state of our earth is concerned can be examined. We know that plants and other animals besides humans can do nothing of their own will because they have no wills of their own. Indeed, they can't add to or take away from the environment we share with them. They live as the Ns factor dictates. Would it be reasonable to imagine that plants and animals that have no wills contribute to our earth's problems? Of course, not. However, human activity can upset the natural balance in ways that involve animals and plants. In this case, the unnatural environments that we create disrupt their biological activities.

One may be tempted to think that the state of our earth is due to, perhaps, the design of our planet is incomplete. When completed at a future date, the misery and unhappiness on this earth will end. That would call to question the knowledge, power, and glory of the Maker. Think of the three examples of His miraculous work described above: the sheer number of things He created, the incredible distances among the things He created, and the complexities of His designs like the animal body systems and the human brain. Is it possible that the Creator of all these things could make mistakes or forget something? No!

His creation was certainly complete because there is nothing in His creation that indicates otherwise. Is it possible that the state of this earth is due to design errors or lack of knowledge on the part of the Designer? The answer Is a capital No! From where are the problems of this earth coming? They are not from a design error or that our earth's design is not complete yet; they are not from animals. Then, we must focus our attention on humans – the part played by Nurture. Then we will reexamine nature in more detail later.

Scientists believe, using Mitochondria DNA, that Africa was the original birthplace of all humans, i.e., all humans can trace their origin to the mythical Adam and Eve. It does not take Rocket Science to believe that, as we can go forward in time from parent to the child, to the grandchild, to the great-grandchild, to great-great-grandchild, and so on; that we can also trace backward, starting from any point in time. We can start from the child to the parents, then to the grandparents to the great grandparents to the great- great grandparents and finally to 'Adam and Eve.'

They bolstered their argument by pointing to the fact that during the hunting and gathering stages in human evolution, here on earth, there was a lot of wandering. Yes! There was a lot of wandering from place to place because they were in constant motion since they must look for food to eat.

When the food (mainly fruits, vegetables, fish, and wild games) was exhausted in a location, they had to move to a new site where the food was available. Remember, the Creator had planted the need to eat. Earlier peoples believed that the earth was much smaller than its current size. So, after many centuries of wandering, humans found themselves in Europe, Asia, etc. They also believed that skin tones, hair types, and physical structures were due to environmental changes.

So, initially, God did not create Chinese, but they were Africans that wandered into China? That also will mean that the distinctive facial structure of a group of people, Japanese, was due to environmental changes over the millennia. The question to this group of scientists is, "Will the facial structure, skin tones, and hair types of a group of West African families taken to China, allowed to live together and marry among themselves, change into Chinese after millions of years?" Of course, the answer is no. The truth is that if a group of West Africans lives in North America, Europe, or Asia, they and their descendants will remain Africans, barring interracial mixing. They will not change into white, red, etc., because of the new environment. There are no examples found in nature.

West Africans forced out of their homelands and sent to different parts of the world remain dark-skinned even after many centuries of living in those foreign countries. Specific characteristics like behavior, mode of processing information, lifestyle, etc., will change, of course. These qualities are external, acquired with time, and from the prevailing environment. But the original self will remain. Right from the beginning, God created Africans as Africans and placed them in their assigned region of the earth with their distinctive features and cultures, as it is true with Chinese, Japanese, Indians, etc.

One can also argue that, initially, Africans wandered into India, all their African characteristics changed, but their skin tone remained? One could even say that the Indians have initially been British people (not Africans). But on wandering into India, their skin color turned dark due to environmental changes. That would not make sense! The Creator created them Indians, not Africans, right from time. There is nothing in our physical world that can prove otherwise.

The above discourse brings up the story of Adam and Eve, the assumed foreparents of all humanity. Would it be reasonable to believe, factoring in the infinity characteristics of the Creator, that He created only one 'Adam and one Eve'? There is nothing in the physical world that shows that God created that way. Modern scientists have discovered that there is not one but many universes. They even question if the earth is the only planet that contains life.

Nothing is one, only the Creator!

Then the story of Adam and Eve, including the story of the 'Devil' appearing as a snake that convinced them to disobey the Creator, could not be genuine! God does not create in ones but multiples.

It would be helpful to take a second look at the facial structure of a Nigerian and that of a Chinese. You will not see an atom of resemblance between the two, except that both are humans with eyes, noses, mouths, etc.

These groups must have had their own 'Adams and Eves'. How can we be confident of this? For starters, the Creator is the God of infinite variety. Nothing created is only one type, size, color, etc. Nothing! Only He is one, only one Creator. Why would He create one Adam and one Eve?

Physically, the Adam and Eve Chinese descended from could not be the same as the Indians descended from or Japanese or American Indians. And the story that goes with this myth is the mere human imagination, particularly males—the weaker (female) is always blamed and made the scapegoat.

Every culture has a 'creation' story to tell, which is different from any other. Think about that! If you think this is a mere hypothesis, test it on friends from different countries.

When people of Southern Nigeria (Easterners and Westerners together) with distinctive cultures are compared, it is easy to see that what separated the peoples of this region was time and distance. If you were to trace back, it would be easy to find out that they descended from the same Adam and Eve. But different physical features can be seen when comparing these groups with the Fulani's and the Hausas in the Northern part of that country.

For centuries, these people were trading with the peoples of Arabia. Consequently, racial mixing occurred through marriage. This history can explain why some Northerners bear the same physical features as Southerners while some (not a majority) don't. In summary, some Northerners are different from the Southerners due to time and distance. But both share the same 'Adam and Eve,' but the rest due to time, distance, and interracial mixing with the Arabian neighbors don't.

At this point, I will discuss Migration and Adaptation. When God created living things, He assigned them to their places. Fishes must live in water and wild animals in the forests. The Creator equipped animals with protective coloration that camouflages them in their environment. A Ghanaian is created to live in a warm and sunny place, now called Ghana.

People assert that first humans moved from one place to another during the wandering and gathering stages of human evolution here on earth. In other words, there was no assignment of locations by the Creator.

Could it be true that fishes were living on land (or somewhere else) initially but wandered into the waters of the world and made their homes there? Or that the Europeans wandered into equatorial regions of the world during the wandering periods and later made the places their homes? Then over time, skin pigmentation and physical structures changed. Nothing in nature can point to the validity of these postulations.

As discussed before, people believed that the first humans started in Africa. But some of them moved away to other places in the world during the wandering stage of human development. So, Chinese, Europeans, Asians started life in Africa but wandered out to their current homes. Then after millions of years and environmental factors, their structures, including facial and hair types, color, etc., changed.

Was it possible they carried with them, from Africa, all their food types, crops, plants, vegetables, oil, and animals which, due to time and place, changed to what exists in these places today? No! These things were there right from the beginning as designed by the Creator. Of course, you can find some non-native stuff in a place. We know people bring back home things they admire or are attracted to from places they visited. But natives are easy to identify.

This postulation suggests that the Creator had no plans for the settlement of humans. He created them, gave them free will, and told them that they were on their own to find a suitable location on earth to occupy. We know that the Creator of all things, with infinite power and knowledge, does not leave His creations hanging.

We are talking about the Creator of all things visible and invisible. He was the Universe designer and hanging them in space without physical supports, and never had a record of collision. He was the designer of the most complex machine, the human body, including the brain. All the cells in human and animal bodies, millions of them, He takes care of each. He can't make a mistake or forget a thing.

Also, the body of an Eskimo is designed to live in cold regions of the world. They have their type of food, vegetables, and fruits. The Creator creates their body to accept the food items found in their kind of environment.

There is a school of thought that talks about the ability of the body to adapt. Adaptation, in that sense, means a 'temporary fix.' If a Chinese finds himself in Congo, for example, he is supposed to stay for a short time for business or vacation. The body can tolerate or endure the new temporary inconveniences. But not forever, since the Creator did not design the body for that environment. Nobody should say that a British, for example, cannot typically adapt to British food or British culture. No! The food or the culture is in the DNA as the Creator made it.

The word 'adaptation' when referring to any living thing in its natural environment is a misnomer. In that context, the implication is that the organism is alien to the environment. As such, the organism had to develop new behaviors or grow new features to help it cope with its new environment.

There is nowhere in nature where an organism behaves or feels like an alien in its environment. Animals in cold regions grow thicker fur or add more fat under their skins – a way of life. They did not, all-of-a-sudden, find themselves in a cold region and had to scramble for thicker fur to ward off the unexpected cold.

The cold region is their home, where the Creator put them. That condition had been that way since time began. But, suppose a Kenyan had to migrate to Russia for some reason? There, everything is new and different from his everyday experience. Whatever he did to cope with the unique situation is a good description of 'adaptation.' He must be creative in finding ways to cope.

American Indians can't see themselves as trying to adapt to American Indian food. The food items in their locality are the type they know, grew up with, and eat. It would take time for them to get used to any other kind. As stated before, when people use the word 'adaptation,' what comes to mind is a temporary fix. And a 'temporary fix' implies 'lasting for a short time.' The next question is, 'How long is a short time'? In the context of adaptation, it means 'as long as' whatever brought about the need for the adjustment lasts but in decreasing order.

Carbohydrates, no matter from which crops or region of the world, are the same and performs the same functions in the body. The characteristics and the nutrients they supply to the body are also the same. It shows that the duties of humans and animals remain the same no matter where they are in the world.

A Zulu person living in Israel must adapt to the food in his new home since it will supply the same nutrients, giving him no reason to diminish his God-assigned duties. Again, the responsibilities of an 'adult' living thing are: "go get" your food, eat it; protect yourself and what is yours from predators and make sure your type exists here forever; and to humans only, He gave an extra function – to take care of the earth.

Chapter 9

SELF-PRESERVATION

Consider the statement, "Protect yourself." How are we sure the Creator wills that all things must protect themselves? Of course, we can't go to him to ask the question or receive the answer. There is no other choice but, again, to examine what He created.

To all His creatures, He provided some means of self-protection. Examine all animals; with no exception, each has a way to protect itself and, in most cases, a way to fight back. Humans have intuition, intellect, instinct, and some physical strength to fight back or protect themselves. Some animals may have tough skins making it difficult for a predator to bite through. Some may possess agility, protective coloration, horns, hoofs, and sharp canines to fight back. Snakes and scorpions have poisonous stings.

He provided the means for protection to all His creatures. He knew the need for self-protection would arise. Did He intentionally build danger like creating predators into the ecosystem? When a country acquires arms and trains its soldiers, war is imminent, or there is preparation for one just in case. Let's dig deeper.

It is almost automatic for animals, if necessary, to fight to protect or defend themselves or what belongs to them. Humans were given the intellect for this purpose and conscience to control/temper excessiveness while other animals have the instinct. Other animals don't need a conscience. Why? They don't have wills of their own. The lion does not go on a killing spree. Nature wired him to take just enough and no more. He could kill when protecting itself. In rare cases, when animals act outside of these natural boundaries, they are considered mad. Generally, animals can't function outside of the dictates of the Creator.

Think of a child's reaction when somebody tries to take away or break their toys or a man's behavior when his child or wife is in some danger. We know the response of a mother when the nourishment, comfort, or safety of her child is in jeopardy. It would be best if you coaxed the mother first before attempting to touch a puppy or chicks. It will be suicidal to try to harm or even touch a lion's cub for the fun of it.

The Creator provided protective coloration to many animals to confuse predators. The most wondrous is the type given to chameleons. Even some animals feign dead when confronted by a predator. Some animals are very cunning, while some animals like lions, fishes, etc., use agility to escape danger.

The Creator provided armor, i.e., hard shells to some animals like crocodiles, turtle, snails, etc., to avoid harm. Imagine a predator trying to swallow a turtle! Don't forget the poisonous stings of bees, snakes, wasps, scorpions, etc. They use these dangerous stings to fight off danger. Bees don't go about looking for an animal to sting. It does so when it senses danger. Some animals will avoid being eaten simply by oozing out some horrible and unbearable odors like the skunk.

Plants protect themselves differently since they are not equipped to move about; they don't possess hands or the ability to shout for help at the sight of danger. Some plants have stinging and or poisonous liquids. Poison Oaks will cause a painful and itching sensation to any person that goes too close to it. Some plants like the rose or trees like the orange, oil palms, etc., are provided with thorns to prevent predators from eating or destroying their fruits before the fruits are ready or ripe to provide the nutrients the Maker demands of them.

Remember, plants like animals don't have wills. If you collect vegetables from a plant or the tree trunk is cut down for whatever reason, there will be no resistance or outcry from the plant or tree. They have no wills of their own. They will readily give up part or all of themselves in obedience to the Creator.

But as a reward, God transferred the burden of self-preservation to their 'enemies,' animals. How? The Creator provided the nutrients, medicines, services, etc., essential for preserving human and animal existence through them. These other living things can't survive here on earth without the food, medicines, protection, and services, etc., these plants and trees provide. Therefore animals, particularly humans, must take care of the plants and trees for their survival, making it unnecessary the need for hands and legs for fight or flight. Thus they are valued, fully protected.

But carnivorous animals, including humans, must kill and eat other living things to survive? To avoid being eaten and exist here on earth explains the Maker's reasons for survival mechanisms for animals and plants. Fighting may not be to get food only. Animals fight for self-defense, to protect their homes and property, plus the security of their families. These activities must happen: It is their way of life. Hence the All-Knowing Maker provided defense mechanisms to assure their survival.

Why would harmless animals that willingly submit to butchering without resistance, animals like cows, goats, etc., need a defense mechanism? These animals, generally, are so many times bigger and stronger than humans. Think about that: If cows decide to fight back when being sent to the butcher, there will be much bloodshed. But they don't, not to human beings! Even the most harmless animals fight for the same reasons enumerated above. God is God; He knows everything.

The humans were given the responsibility to protect them from other animals. We now know that the Maker intends to have these animals killed but for food and other services. To make sure animals and other living things are protected, humans are given the intellect and conscience to make laws that protect them. Killing animals for the fun of it is prohibited in all societies. So, the protective 'weapons' given to the harmless animals are for fighting off other animals to ensure their survival.

Wild animals do not obey human laws but the Creator's. As such, these animals will kill off themselves if not restricted. So, by providing the defense mechanism, the Creator made it extremely difficult for animals to go on a senseless killing spree.

No animal kills its kind for food. But animals do fight one another to protect their food or space or to protect their belongings. And humans know better not to kill off all animals or destroy vegetation because their well-being depends on the food, medicines, and services these things provide.

At this point and based on observation, I will propose an answer to the question about whether God intentionally created insecurity in living things' ecosystems. To understand why the Creator so-equipped these animals (including humans), we need time to read His instructions written in His creations – the Divine recycling and the Perpetual cycle systems. One should remember that there is no new water, air, animals, plants, etc., created.

As it is ordained, if a lion, for example, needs some food, he knows to 'get' or go and hunt and kill an animal to satisfy the hunger. With other predatory animals like humans, similar events will occur. To meet the protein requirements of our diets, we look for some fish or animal meat. In other words, we will have to kill some fish or animals. Plants 'suffer' similar faith when humans need some vitamins or medicines the Creator packed in these plants.

In all these events, we restrict ourselves to taking what is required to satisfy the need. The restriction is a Divine Law. Humans generally flout this law, thus incurring the consequences – the sufferings we experience here on earth.

What is going on here is the Divine Perpetual Cycle (see fig. 6), including Divine Recycling. If we take or kill only what we need, the number of animals or plants will not be affected due to the Perpetual Cycle. The killing was for a practical purpose – to keep and maintain the living.

So, the defense mechanism provided to these animals is only a deterrent that discourages or frustrates excessiveness on the strong on the weak and helps keep the Creator's recycling system and life on earth to continue forever.

Another source of strife is immigration to a foreign land or forcibly removing a living organism from its environment into an alien one! Is there anything in nature that shows the Maker's intention for an organism was to abandon its environment to live in another? No!

Let's examine some scenarios: Some people adopt birds as pets. They provide all the birds' needs, including special cages that offer maximum comfort. What happens on the day the cell was mistakenly left open? The bird flies away, never to return! The bird›s reaction shows it was not the Creator›s intention for birds to live in human homes, in cages.

We can go further to examine the cases of adopted children. People adopt children in different circumstances. In all situations, only orphaned children cope the best. "Cope!" If treated lovingly and fairly, he grew up grateful to the new family to come to his rescue when the natural parent was unavailable.

But a child that was abandoned or given up because of poverty will seldom grow up a satisfied or contented person, no matter how loving and caring the new families were. The child will always long for the natural parents. At the back of his mind will always be that feeling of loss and not belonging to the new family. It is not the fault of the new family or the child. It is simply not the intention of the Maker.

A child or any person belongs to the family the Creator assigned him to. Compare this with the relationship dogs have with humans. A visit to shelters housing stray or abused dogs will best illustrate this point. As soon as you enter the facility, you will be greeted by the howling of the dogs, almost 'begging' to be adopted.

Take one home; the dog feels at home. Leave the dog in the house and travel. The dog will be there waiting; it will not run away. Does the behavior not show that the Maker designed it to be so? Your home or environment is where the Designer chose for you and furnished it with all your body will need. Changing what God designed will only cause suffering and discomfort. Dogs and a few other animals have a closer relationship with humans and are designed to live anywhere on this earth.

Finally, in the case of Jews who were forced out of their environment and scattered all over the globe and concentrating in Germany or Africans who were forcefully removed from their homes and resettled in the Americas during slavery, the feeling is the same. The same goes for even Africans who were self-exiled to Europe, America, etc., due to unacceptable conditions in their home countries; the feelings are the same. There is always the feeling of being a foreigner or subculture in an Alien Country. Even the host countries will always see them as aliens. There is no scenario where this is not the case. It does not matter who is involved – Ghanaian, British, Swiss, Italian, Egyptian, etc. Doesn't this show that the Creator never intended that any organism live in any other environment, but at the place, he assigned to it? Of course, the forced removal of any creature from its environment will always violate God's laws. Consequently, incurring sorrow.

The urge or the desire of an adopted child to seek out the biological parents; the desire of the Africans in America and Europe to go back to Africa or the Jews to have a place they call home is incredible. In some cases, the new homes of these peoples or even animals are better than the ones they left behind. But the longing will always exist. Many Nigerians living abroad come from villages much less developed than their current homes in Europe and America. But the majority or almost all would still prefer to go back to their original homes. The longing is natural and should not be surprising!

When God created all living things, He did not ask them to choose the places in which they liked to live. No! He did the choosing for them and provided food suitable for their bodies there. Being the Designer, He knew what food their bodies needed and put all the needs in their locations.

I have already explained why God gave dark skins to black people or why some animals can see in the night, but some can't, and why yams grow in equatorial regions of the world. God is All-Knowing. He can't forget a thing. There is a reason for anything He Created and positioned in the universe.

When an automotive engineer designs the car, he knows the type of fuel/gas his invention would need for power and the terrains for which it is suited. God created the human 'machine.' He knew the environment that best suits it and fuels its needs.

Do humans face more daunting social and cultural problems in a foreign country than other animals or living things? No! In this respect, all organisms are similar. The only difference is humans have ways to voice their objections. How will a bird whose owner cages in a beautiful home tell of its displeasure to be a prisoner in the house? What of a tropical plant uprooted from its natural environment and replanted in a desert? Indeed, it will not be happy there, and it is likely to survive only with more care.

Indeed, the earth's caretakers need to take a breather, take time out to go back to nature to understand why, notwithstanding the advances in science and technology, the earth is still not the Paradise it was created to be. The design of our planet is complete. We are sure there were no design errors because the All-Knowing, the All-powerful, created our earth. And we know that the earth was created to be a happy abode for all living things. Humans need a time out to think!

Chapter 10

THE FAMILY

As described earlier, a lot of comparing is involved when making choices. Some of these negotiations may lead to disagreements and discontentment. We often forget that anything God created is unique, an entity different from any other, even from its kind and for a particular purpose.

Consider the following comparisons: orange fruit and grapefruit, which is better? What of pea and mango or lemon fruit and tangerine? What of cow meat and goat meat? What of a Nigerian marrying an Ethiopian woman or a German or a Japanese instead of a Nigerian? Sometimes we even hear of some bizarre questions like, "which is preferred, a father or a mother?" No person can answer questions of this type without more information.

Let's start with choosing between peas and mangoes. A pea is not a mango or vice versa. They don't look alike; neither do they taste the same. So, when comparing or making a choice between the two, for which characteristics will the person doing the choosing be looking at – the color, the size, taste, etc.? They are in no way the same. But if the choice is between two apples or two pineapples, for example, that will be easy: Since they are the same, one may consider their sizes depending on how much one may need. Or one may consider how ripe they are, depending on whether he likes soft or unripe apples.

It's the same dilemma we find ourselves when confronted with more difficult choices like choosing between ones' mother and father. Though humans, the father, and mother are not the same. To make such a choice will require more information.

A typical father is a provider, the pillar of his family, a source of protection, security, and safety for the family. When 'the father' is around, there is generally a feeling of security. If there is no father in a family, the children, including the mother, will always be missing an important member of the family.

As described earlier, a lot of comparing is involved when making choices. Some of these negotiations may lead to disagreements and discontentment. We often forget that anything God created is unique, an entity different from any other, even from its kind and for a particular purpose.

Consider the following comparisons: orange fruit and grapefruit; which is better? What of pea and mango or lemon fruit and tangerine? Can you even compare cow meat and goat meat? Some people consider a Nigerian marrying an Ethiopian woman or a German or a Japanese instead of a Nigerian? Sometimes we even hear some bizarre questions like, "which is preferred, a father or a mother?" No person can answer questions of this type without more information.

Let's start with choosing between peas and mangoes. A pea is not a mango or vice versa. They don't look alike; neither do they taste the same. So, when comparing or choosing between the two, what characteristics will the person choosing be looking at – the color, the size, taste, etc.? They are in no way the same. But if the choice is between two apples or two pineapples, for example, that will be easy: Since they are the same, one may consider their sizes depending on how much one may need. Or one may think how ripe they are, depending on whether he likes soft or unripe apples.

It's the same dilemma we find ourselves when confronted with more complex choices like choosing between ones' mother and father. Though humans, the father and mother are not the same. To make such a choice will require more information.

A typical father is a provider, the pillar of his family, a source of protection, security, and safety for the family. When 'the father' is around, there is generally a feeling of security.

In a fatherless family, the children, including the mother, will always be missing an essential member of the family.

It is also difficult for a typical family to make it without the mother. A typical mother is a wife first, then a mother, the selfless caregiver, the first teacher of the child, the family counselor. Unfortunately, in a one-parent family, the available parent plays the roles of both the father and mother. Can a mother play the part of the father of a family? No, and neither can the father play the role of a mother – the Creator did not design it so. Raising a family is a big job for one person: a mother has a lot on her plate; so also, the father! The following explains the point I'm trying to make.

One day I visited my friend to accompany him and his family to a concert. Immediately I stepped into the living room; I could feel an atmosphere of urgency because of my presence. Why? Everybody knew the event would start in one hour while the driving distance would take forty-five minutes. But the family was not ready yet.

The first scream I heard came from the early-teen daughter who announced that her life was over because she could not find her new pair of shoes and appealed to the mother for help. The poor woman, about to put on her clothes, had no choice but to run upstairs to help the grieving daughter.

On her way to the daughter, she met her two younger sons engaged in a fistfight about who should be the first to brush his teeth, "as daddy had instructed." The mother settled the 'case' by advising and insisting that the person that entered the bathroom first should brush first. Then she made it to the daughter, who had started crying. Finally, they found the pair of shoes. The crying stopped only after the mother promised to take her to the mall later.

Her 'sweet' job was not over yet because the husband was having trouble putting on his ties appropriately and was waiting patiently for the wife's help. It was about ten minutes to their departure time; the mother barely found time to get herself ready. We all got into the car and waited for her. One of the children complained, "Mom is always late." Immediately she came out of the house, went straight to the husband, kissed him, took her seat, and said nothing.

That's a wife and mother, always taking care of the needs of her family! Before I forget: the woman I'm talking about, any 'mother,' have you ever tried to lift her handbag? The least weight I have lifted was about five pounds.

I'm almost sure you have experienced or observed the following scenarios: "Mother, I'm thirsty." The mother pulls a small bottle of water or soda from her handbag and hands it to the child. "Honey, I'm sweating, but I forgot my hankie." "Honey" pulls a napkin or paper towel from the same handbag for the husband. Don't forget that somewhere in the same purse are her makeup, personal things, and much more. I'm sure you get the picture. You wonder how she was ever so prepared and ready for all the eventualities! "Could this be a learned behavior," we are tempted to ask? No! It is because the Maker had already planted the abilities to develop that 'caring' spirit in her. People will always describe such a woman as a mother! Let's dig deeper!

I once encountered a man whose wife had given birth to a bouncing baby girl. There was a heated argument about how the baby could be cared for when the wife returned to work. The man made two suggestions: Since the two had jobs, they should employ a babysitter. Alternatively, he could become a stay-at-home dad and take care of the baby. And the wife could continue her better-paid job.

The wife could not accept any of those suggestions. She took issue with employing a stranger to care for her newborn for even an hour or a day. She could not understand how any mother, under similar circumstances, could leave her baby to make the extra money to buy 'stuff.' She understood why a woman who didn't have the choice she had could do that. She felt it was morally wrong and unfair to the baby. Then she declared she had enough work to do in her house or could work part-time from home.

She wondered how her husband or any husband, can take care of a baby in any circumstances, considering the type of temperament men had. "The nature of men did not support that idea," she argued. That's a 'mother' talking, one of those women who would like to turn this earth into heaven, a woman utterly devoted to the happiness of her family. From this scenario, it is evident why not all women can qualify for the two titles – 'wife' and 'mother.'

Motherhood or fatherhood is a divine assignment and exemplified by God-fearing couples. But in few families, you may meet a mother or wife who feels the divine-assigned work is beneath her. She may even believe nature cheated her for making her a female. Such a belief is one source of the unhappiness we experience here on earth.

The Creator God cannot make a mistake, and His creation, designed to the minutest details. As the mythical story has it, "God saw Adam (any Adam) lonely and sad and knew he could not make it without a comforter, a helper a soul met. He then created a mate, wife, mother, for him for that purpose. As a result, Adam's world transformed into a paradise. It is likely Adam thankfully said, "I am now complete." The two became one and inseparable.

It is clear what was the intention of the Creator in designing Adam and making Eve the spouse. We are sure that marriage is an institution God Himself established so that all animals would have a mate. It is said, "What God put together, no man should put asunder."

That marriage is a phenomenon becomes undeniably clear when we consider the tremendous difficulties families that have the misfortune of having only one of two partners. Also, it is difficult for a man to live without a wife or vice versa. It is likely you have experienced the soothing touch or hug from a mother or wife to a grieving husband. What of the heart-warming or reassuring smiles from a wife and mother to the husband returning from his job? It is simply miraculous. It is divine!

Let's go back to nature and consider how other animals raise a family and care for their young ones. Remember, goats, dogs, lions, fishes, worms, etc., raise families in their ways. Chicks and other young ones will stand almost immediately after hatching and start eating and are not breastfed. These include birds, fishes, and many reptiles like crocodiles, snakes, etc. But the young ones of goats, lions, dogs and other animals that give live birth, may take some few hours before their young can stand and expect food from the parents. Most are breastfed. The young ones will take about eighteen months of nursing and rearing before they can become independent. It is in very few cases in nature where rearing the young is the duty of the male.

Human babies are an entirely different story. They are the most helpless young ones in the animal kingdom. Only with the help of the mother can they even eat, mainly breastfeeding or regular milk. It takes them about one year to sit, talk, and walk. For about eighteen years, will they be at the parents› care, protection, and tutelage (this age varies among cultures).

Why would humans' children take about eighteen years to reach maturity, but other animals take about eighteen months? The best answer must come from studying nature around us – what sets humans apart from other living things. It is the gift or possession of the intellect, the senses, the conscience, and the Creator's duties assigned to them.

The development and nurturing of the intellect and the senses take time and start from childbirth. Some people put it as early as conception. Evidence? Some parents start teaching their children – math, science, and technology, music, etc., when the children were still in the womb. Development and expansion of the brain continue past the point of wisdom to ripe old age. From this point, intellect starts to decline and continues until death extinguishes the flame of life.

Children will eventually become the earth's future caretakers, and the parents must take enough time to raise them well. Unlike other living things, raising the future caretakers will not be an easy task and will require the full attention and cooperation of both father and mother, not one parent. Any person that has raised a family knows that one parent will not do. But with other living things, one parent may do. For this reason, marriage is made a binding contract from the Creator Himself. The consequence is too great for an unnecessary or thoughtless dissolution of a marriage.

But our "civilized" modern societies think differently or believe they can do better than the Creator to the extent that they now see marriage as a non-binding Contract between two people, an arrangement they can dissolve at will for flimsy excuses. Modern societies have established different norms for marriage, including laws that guide them. In some corners of the earth, justifications are not required to dissolve a marriage. All a person needs to do is say the word 'divorce' to the judge, and he/she grants the request.

In such societies, many unnatural behaviors and practices are commonplace. It is alright for a mother to leave her few weeks and, in some cases, 'few days' old babies to return to work when the baby-mother bonding had not even taken place, or the birth- wounds of the mother, not healed. It is okay for a man and a woman to live together without getting married. The child doesn't have to be obedient and respectful to the parents.

How can a child that does not show respect to the parents show respect to other adults, such as the teacher, or subject himself to the molding of society? Adults are not allowed to discipline their children, let alone others. The so-called new way of raising a family is contrary to the very reason why the All-Knowing God put adults in charge of training and caring for children.

There is a famous saying in Nigeria, "it takes a village to raise a child." Even some holy books have the warning, "spare the rod and spoil the child." Remember that the children's cognitions are not as functional as adults'. In some households, nobody is in charge or the head of the family. All is because of a selfish, disrespectful, and domineering partner who neither fears God nor respects His laws. Legal cases against abusive parents have set the norms for all families, unfortunately.

To be respected, though, the man should understand his duties and obligations in the family, not the abusive type that prefers alcohol, smoking, and drugs to his family. A father and husband can't be the type that can blow his whole paycheck gambling when his newborn baby has no milk and diapers. If their livelihood depends on how hard they work (especially physically, such as farm laborers), he should not be the lazy type that leaves all the burden to his wife. And, he cannot be the man only interested in making babies but disappears when the responsibility of providing for the babies arises.

Nothing in nature shows that misunderstanding and quarreling will not be a part of living things' interaction with each other. These are observed among animals that live in the air, land, and water. It cannot be different for humans. These can't be allowed to stand in the way of marriage, an institution the Creator Himself ordained, or cause children, the future earth's caretaker, to be raised by a single parent.

If humans appreciate who the Creator really is and that He established marriage for the tremendous and trying work of raising the future earth's caretakers, divorce would be nearly absent from human interactions. How? If humans know that God is the loving Creator that provided all they need and that without Him, they will not even exist, then out of respect and fear of displeasing a loving Creator, they will obey God's law on marriage.

If couples love and respect each other, understand the importance of the Creator's assigned duties, and above all, are their "brother's keeper," divorce would be rare. It will be easy to forgive. Indeed, humans were not taught right about who the Creator really was or were rather deceived by their religious teachers. Humans must take time out to go back to nature and read God's laws, boldly written in all His creations.

A husband and father - the man, working according to the laws of God - is the man that wakes up every morning, conscious of his obligations to his family, the earth, and himself. He goes to his work (any legal job) every day and returns home to be with his family at the end of the workday. A husband and a wife must show love and kindness toward each other and work together for the family's sake.

Nobody who has been on this path before would dispute that raising a family is not easy. It has its trials and tribulations but also joy and great feelings of worthy accomplishments. The consequences of not following God's laws are the breakup of families and the sadness we see throughout the land. Yet, we ask why the 'unhappiness' we experience every day? The heartbreaks will continue until the transgressions stop.

The following is a familiar story: Two houses were standing next to each other. One of them resided in a miserable family. The spouses yelled at each other; they fought and quarreled all the time. The other was a place of happiness and calm. During one of the fights, the wife asked her husband: "Did you ever hear them quarrel? "No", he answered! "So, go there and see what they do to avoid quarreling!"

The husband stood at the window of his neighbor's house and watched. There, they were busy doing their own thing. The wife was cooking, and the man sat at the table, preparing their tax returns. The doorbell rang, and the man jumped up and headed to the hallway to answer the door. On his way, he bumped into a flowerpot, and it fell and broke. He got down on his knees and started picking up the pieces.

The wife (on hearing the cracking noise) knew something broke and ran into the hallway from the kitchen. On seeing the husband picking up the pieces, she also knelt and started helping her husband. The man said to his wife: "I am so sorry. I rushed to get the door and bumped into the flowerpot. It fell and got broken." The wife replied: "No, dear, it was my fault. I put it there on the way. That's why you bumped into it." They gave each other that unique and loving look and that was over. Then, they went back to what they were doing.

The man standing by the window returned to his wife. She wanted to know what the secret of their happiness was. He told his wife: "I know it now. In their family, they both take responsibility for things that go wrong, but in our family, both of us are always right!" That's the secret of family happiness!

Food for thought: The "trick" isn't taking responsibility alone, but preferably is not always claiming the right. In relationships, there are times when you need to forfeit your right to win the peace: depending on which you cherish most. This recipe is applicable in many situations and all relationships. Ego stands in the way of many good things.

The excellent relationship between a wife and her husband is critical in fulfilling the third function of all living things, particularly humans here on earth – procreation, which will keep their kind here on earth permanent and forever. Because other animals, including plants, have no wills of their own, they obey this command without questions. With humans, it is a different story. Humans can decide not to have children; some may even give up their children for adoption for reasons less than selfishness.

Remember, all living things are commanded to live here on earth, where everything they need to survive is pro d; procreate to extend their existence here on earth forever and to humans only, to take care of the earth. Nothing on this earth shows that procreation is a matter of choice. No li thing other than humans have that choice (but with severe consequences). It is universal and, therefore, the Creator's law.

There are people who (through choice or action of their own) don't have the gift of procreation. Those peoples are exceptions, not the rule. Though, all living things are tied together in a perpetual bond. But procreation is at the top of the ticket.

Think of what will happen to other living things in the absence of human beings on this earth? The Creator knew that our earth would need a caretaker. He assigned that function to humans in addition to maintaining their presence on earth forever.

The offspring of humans, being the future earth's caretakers, must be raised right. Hence, children are left in the care of their parents until they become adults. But is this not true with all animals like, say, goats, cows, worms? What makes the difference is the possession of the brains—intellect and conscience by humans. The raising of a human child, a future earth caretaker, can't be a random act but deliberate, a feat that requires both parents and unity of purpose to accomplish.

The father and mother can't face different directions in this task. It is a trying journey that will stretch patience to the limit, a test of love, commitment, and above all, obedience to the law of the Creator. Many families do their best at this assignment. But few fail woefully, generally due to selfishness and particularly, love of money, lack of love and commitment to the family, insincerity, and skewed relationships. In short, the man and woman must be evenly yoked. When a person 'loves,' he or she must love everything about the person, not 'half-and-half.'

Raising children or family cannot be a contract but a duty, a duty nurtured by love and belief in God. Yes! Think of a situation devoid of love and the acceptance of God's laws? Such families will not be suitable to raise and nurture the next generation of the earth's caretakers. The problem starts when a marriage or union is not conceived in love, with a clear conscience and respect for God's laws. Included in this is the baggage brought into the relationship.

It is also difficult for a typical family to make it without the mother. A typical mother is a wife first, then a mother, the selfless caregiver, the first teacher of the child, the family counselor. Unfortunately, in the absence of one of the two, in some families, the available parent plays the roles of both the father and mother. Can a mother alone play the role of the father of a family? No, and neither can the father alone play the role of a mother – the Creator did not design it so. The raising of a family is a big job for one person: a mother has a lot on her plate; so also, the father!

One day I visited my friend to accompany him and his family to a concert. Immediately I stepped into the living room, I could feel an atmosphere of urgency because of my presence. Everybody knew the event would start in one hour while the driving distance would take forty-five minutes.

The first scream I heard came from the early-teen daughter who announced that her life was over because she could not find her new pair of shoes and appealed to the mother for help. The poor woman, about to put on her clothes, had no choice but to run up-stairs to help the grieving daughter.

On her way to the daughter, she met her two younger sons engaged in a fistfight about who should be the first to brush his teeth, "as daddy had instructed." The mother settled the 'case' by advising and insisting that the person that entered the bathroom first should brush first. Then she made it to the daughter who had started crying. Finally, they found the pair of shoes. The crying stopped only after the mother promised to take her to the mall later.

Her 'sweet' job was not over yet because the husband was having trouble putting on his ties properly and was waiting patiently for the wife's help. It was about ten minutes to their departure time; the mother barely found time to get herself ready. We all got into the car and waited for her. One of the children complained, "Mom is always late." Immediately she came out of the house, she went straight to the husband and kissed him and took her seat and said nothing.

That's a wife and mother, always taking care of the needs of her family! Before I forget: the woman I'm talking about, any 'mother,' "Have you ever tried to lift her handbag?" The least in weight I have lifted was about five pounds.

I'm almost sure you have experienced or observed the following scenarios: "Mother, I'm thirsty." The mother pulls a small bottle of water or soda from her handbag and hands it to the child. "Honey, I'm sweating, but I forgot my hankie." "Honey" pulls a napkin or paper towel from the same handbag for the husband. Don't forget, somewhere in the same handbag are her make-up, personal things, and much more. I'm sure you get the picture. You wonder how she was ever so prepared and ready for all the eventualities! "Could this be a learned behavior," we are tempted to ask? No! It is because the Maker had already planted the abilities to develop that 'caring' spirit in her. People will always describe such a woman as a mother!

I once encountered a man whose wife had given birth to a bouncing baby girl. The couple was having a heated argument about how the baby could be cared for when the wife returned to work. The man made two suggestions: Since the two had jobs, they should employ a babysitter, or he could become a stay-at-home dad and take care of the baby while the wife continued her job that fetched more money than his.

The wife could not accept any of those suggestions. The suggestion of employing another person to care for her baby for even an hour or a day to allow her to 'work' repulsed her. She could not understand how any mother, under similar circumstances, could leave her baby to make the extra money to buy 'stuff.' She understood why a woman who didn't have the choice she had could do that. She felt it was morally wrong and unfair to the baby. Then she declared she had enough work to do in her house or could work part-time from home.

She wondered how her husband or any husband, can take care of a baby in any circumstances, considering the type of temperament men had. "The nature of men did not support that idea," she argued. That's a 'mother' talking, one of those women who would like to turn this earth into heaven, a woman completely devoted to the happiness of her family. From this scenario, it is evident why not all women can qualify for the two titles – 'wife' and 'mother.'

Motherhood is a divine assignment. Think of a situation where there is no mother or the woman who feels the divine assignment is beneath her or nature cheated her for making her a female? Such a belief is one source of the unhappiness we experience here on earth.

The Creator God cannot make a mistake and His creation, designed to the minutest details. As the mythical story has it, "God saw Adam (any Adam) lonely and dejected and knew he could not make it without a comforter, a helper a soul met. He then created a mate, wife, mother, for him for that purpose. As a result, Adam's world transformed into a paradise. It is likely Adam thankfully said, "I am now complete." The two became one and inseparable.

It is clear what was the intention of the Creator in designing Adam and making Eve the spouse. We are certain that marriage is an institution God Himself established so that all animals would have a mate. It is said, "What God put together, no man should put asunder."

That marriage is a phenomenon becomes undeniably clear when we consider the tremendous difficulties families that have the misfortune of having only one of two partners. Also, it is difficult for a man to live without a wife or vice versa. It is likely you have experienced the wondrous and soothing touch or hug from a mother or wife to a grieving husband. What of the heart-warming or reassuring smiles from a wife and mother to the husband returning from his job? It is simply miraculous. It is divine!

Let's go back to nature and consider how other animals raise a family and care for their young ones. Remember, goats, dogs, lions, fishes, worms, etc. raise families in their ways. Chicks and other young ones will stand almost immediately after hatching and start eating and are not breast-fed. These include birds, fishes, and many reptiles like crocodiles, snakes, etc. But the young ones of goats, lions, dogs and other animals that give live birth, may take some few hours before their young can stand and expect food from the parents. Most are breast-fed. The young ones will take about eighteen months of nursing and rearing before they can be on their own. It is in very few cases in nature where rearing the young is the duty of the male.

Human babies are an entirely different story. They are the most helpless young ones in the animal kingdom. Only with the help of the mother can they even eat, mainly breastfeeding or regular milk. It takes them about one year to sit, talk, and walk. For about eighteen years, will they be at the care, protection, and tutelage of the parents (this age varies among cultures).

Why would humans' children take about eighteen years to reach maturity, but other animals take about eighteen months? The best answer must come from studying nature around us – what sets humans apart from other living things. It is the gift or possession of the Intellect, the senses, the conscience, and the duties the Creator assigned to them.

The development and nurturing of the intellect and the senses take time and start from childbirth. Some people put it as early as conception. Evidence? Some parents start teaching their children – math, science, and technology, music, etc. when the children were still in the womb. Development and expansion of the brain continue past the point of wisdom to ripe old age. From this point, intellect starts to decline and continues until death extinguishes the flame of life.

Children will eventually become the earth's future caretakers, and enough time must be taken to raise them well. Unlike other living things, the raising of the future caretakers will not be an easy task and will require the full attention and cooperation of both father and mother, not one parent. Any person that has raised a family knows that one parent will not do. But with other living things, one parent may do. For this reason, marriage is made a binding contract from the Creator Himself. The consequence is too great for an unnecessary or thoughtless dissolution of a marriage.

But our "civilized" modern societies think differently or believe they can do better than the Creator to the extent that they now see marriage as a non-binding Contract between two people, an arrangement they can dissolve at will for flimsy excuses. Modern societies have established different arrangements for marriage, including laws that guide them. In some corners of the earth, excuses are not required to dissolve a marriage. All a person needs to do is say the word, 'divorce' to the judge, and he/she grants the request.

In such societies, many unnatural behaviors and practices are commonplace. "It is alright for a mother to leave her few weeks' and in some cases, 'few days' old babies to return to work, when the baby-mother bonding had not even taken place, or the birth- wounds of the mother healed. It is okay for a man and a woman to live together without getting married. The child doesn't have to be obedient and respectful to the parents.

How can a child that does not show respect to the parents show respect to other adults, such as the teacher or subject himself to the molding of the society? Adults are not allowed to discipline their children, let alone others. The so-called new way of raising a family is contrary to the very reason why the All-Knowing God put adults in charge of training and caring for children.

In Nigeria, there is a popular saying, "it takes a village to raise a child." Even some Holy Books, have the warning, "spare the rod and spoil the child." Remember that the children's cognitions are not as functional as adults. In some households, nobody is in charge or the head of the family because of a selfish, disrespectful, and domineering partner who neither fears God nor respects His laws. Legal cases against abusive parents have set the norms for all families, unfortunately.

To be respected, though, the man should understand his duties and obligations in the family, not the abusive type that prefers alcohol, smoking, and drugs to his family. A father and husband can't be the type that can blow his whole paycheck gambling when his newborn baby has no milk and diapers. If their livelihood depends on how hard they work (especially physically, such as farm labors), he should not be the lazy type that leaves all the burden to his wife. And he cannot be the man only interested in making babies but disappears when the responsibility of providing for the babies arises.

There is nothing in nature that shows that misunderstanding and quarreling will not be a part of living things interaction among each other. These are observed among animals that live in the air, land, and water. It cannot be different for humans. These can't be allowed to stand in the way of marriage, an institution the Creator

Himself ordained or cause children, the future earth's caretaker, to be raised by a single parent.

If humans appreciate who the Creator really is and that He established marriage for the tremendous and trying work of raising the future earth's caretakers, divorce would be nearly absent from human interactions. How? If humans know that God is the loving Creator that provided all they need and that without Him, they will not even exist, then out of respect and fear of displeasing a loving Creator, they will obey God's law on marriage.

If couples love and respect each other, understand the importance of the Creator's assigned duties, and above all, are their "brother's keeper," divorce would be rare. It will be easy to forgive. Surely, humans were not taught right about who the Creator really was or were rather deceived by their religious teachers. Humans must take time out to go back to nature and read God's laws, boldly written in all His creations.

A husband and father, the man, working according to the laws of God, is the man that wakes up every morning, conscious of his obligations to his family, the earth and himself, goes to his work (any legal job of his choice), every day and at the end of the workday returns home to be with his family. A husband and a wife must show love and kindness toward each other and work together for the sake of the family.

Nobody who has been on this path before would dispute that raising a family is not an easy task. It has its trials and tribulations but also joy and great feelings of worthy accomplishments. The consequences of not following God's laws are the breakup of families and the sadness we see throughout the land. Yet, we ask why the 'unhappiness' we experience every day? The heartbreaks will continue until the transgressions stop.

The following is a familiar story: Two houses were standing next to each other. One of them resided in a very unhappy family. The spouses yelled at each other; they fought and quarreled all the time. The other was a place of happiness and calm. During one of the fights, the wife asked her husband: "Did you ever hear them quarrel? "No", he answered! "So, go there and see what they do to avoid quarreling!"

The husband stood at the window of his neighbor's house and watched. There, they were busy doing their own thing. The wife was in the kitchen cooking, and the man sat at the table, preparing their tax returns. The doorbell rang, and the man jumped up and headed to the hallway to answer the door. On his way, he bumped into a flowerpot, and it fell and broke. He got down on his knees and started picking up the pieces.

The wife (on hearing the cracking noise) knew something broke and ran into the hallway from the kitchen. On seeing the husband picking up the pieces, she also knelt and started helping her husband. The man said to his wife: "I am so sorry. I rushed to get the door and bumped into the flowerpot. It fell and got broken." The wife replied: "No, dear, it was my fault. I put it there on the way. That's why you bumped into it." They gave each other that special and loving look, and that was over. Then, they went back to what they were doing.

The man standing by the window returned to his wife. She wanted to know what the secret of their happiness was. He told his wife: "I know it now. In their family, they both take responsibility for things that go wrong, but in our family, both of us are always right!" That's the secret of the family happiness!

Food for thought: The "trick" isn't in taking responsibility alone, but preferably is not always claiming right. In relationships, there are times when you need to forfeit your right to win the peace: depending on which you cherish most. This recipe is applicable in many situations and all relationships. Ego stands in the way of many good things.

The excellent relationship between a wife and her husband is critical in fulfilling the third function of all living things, particularly humans here on earth – procreation, which will keep the existence of their kind here on earth permanent and forever. Because other animals, including plants, have no wills of their own, they obey this command without questions. With humans, it is a different story. Humans can decide not to have children; some may even give up their children for adoption for reasons less than selfishness.

Remember, all living th s are commanded to live here on earth, where everything they need to survive is pro d; procreate to extend their existence here on earth forever and to humans only, to t e care o the earth. Nothing on this earth shows that procreation is a matter of choice. No li thing other than humans have that choice (but with severe consequences). It is universal and, therefore, the Creator's law.

There are people who (through choice or action of their own) don't have the gift of procreation. Those peoples are exceptions, not the rule. Though all living things are tied together in a perpetual bond, procreation appears to be at the top of the ticket.

Think of what will happen to other living things in the absence of human beings on this earth? The Creator knew that the Earth would need a caretaker. He assigned that function to humans in addition to maintaining their presence on earth forever.

The offspring of humans, being the future earth's caretakers, must be raised right. Hence, children are left in the care of their parents until they become adults. But is this not true with all animals like say, goats, cows, worms? What makes the difference is the possession of the brains—intellect and conscience by humans. The raising of a human child, a future earth caretaker, can't be a haphazard act but deliberate, a feat that requires both parents and unity of purpose to accomplish.

The father and mother can't face different directions in this task. It is a trying journey that will stretch patience to the limit, a test of love, commitment, and above all, obedience to the law of the Creator. Many families do their best at this assignment. But few fail woefully, generally due to selfishness and particularly, love of money, lack of love and commitment to the family, insincerity, and skewed relationships. In short, the man and woman must be evenly yoked. When a person 'loves,' he or she must love everything about the person, not 'half-and-half.'

Raising children or family cannot be a contract but a duty; a duty nurtured by love and belief in God. Yes! Think of a situation devoid of love and the acceptance of God's laws? Such families will not be suitable to raise and nurture the next generation of the earth's caretakers. The problem starts when a marriage or union is not conceived in love, with a clear conscience, and respect for God's laws. Included in this is the baggage brought into the relationship.

Four main categories of baggage may play a role in the success or demise of the union:

- Permanent Scars—traumatic or unhappy events; wars, disasters, rape, separation, etc.;
- Guilt—murder, shameful behaviors, and indiscretions.

- Interference in Relationships—whisperings, jealousy, gossip, divisiveness
- Personal Problems—mental, physical, medical issues, family responsibilities
Any or a combination of these can cause havoc to an otherwise happy family.

Children are deeply affected when raised in an unhappy home or in a home where one of the parents is absent or unresolved baggage is interfering in the relationship. There are many examples to support this claim. Just look around you. I have always wondered why this state of many broken and unhappy families seemed to be the new normal. I didn't experience this when I was growing up but now as an adult. Then there was not this litany of garbage people bring into the relationship.

What changed? Yes! Many things changed. Now a person can marry and divorce multiple times. People who engage in this type of behavior believe that problems associated with marriage or raising a family can easily be remedied with a divorce, despite the children. It is legal.

When I was growing up, I learned that marriage was a solemn affair, to the extent of involving family members and their friends and all the members of both villages. In many countries, then, marriage could unite kingdoms and help end wars.

In my village, marrying or being allowed to marry was not an easy task – it came with many responsibilities. First, the man had to prove to the family that he had reached the breadwinner status to get married. Second, newly married couples were expected to join the village council and help resolve the families' issues. While the adults had meetings to solve their problems, children played with one another, secured by the presence of fathers, mothers, aunts, and uncles.

Compare that scene to what occurs in situations where a young man and woman go it alone getting married without bothering to inform their parent(s) and neighbors. To whom will such couples go when, inevitably, there is a marital problem? Of course, not the disregarded parents or close relatives, people interested in the welfare of the family but to paid Councilors and Lawyers, people more interested in their paychecks than the success of the marriage. ?

Currently, young women and young men must finish college first before thinking of marriage. Why? The young woman, particularly, wants to be independent and be able to take care of herself in case of divorce. "In the case of divorce!" So, divorce is already built into the relationship even before it starts. By the end of college, the time has taken its toll on the child-bearing age. One should wonder why attending college and getting married can't take place simultaneously.

Some young people would like to play around first, have their time with alcohol, 'testing' as many of the opposite sex as possible. Birth control drugs and devices will keep such young women and men from getting involved in pregnancy and abortion. The disastrous effects of such drugs on the reproductive organs are well documented.

We hear of children born with cocaine found in their system or young children suffering from the effects of second-hand smoking. Habits acquired during bachelorhood will not disappear simply because one got married. In such a modern family, no one is happy. Products of such families we find in our schools and colleges. They influence other children. Adults from such families are found on our highways driving, working in banks, schools, are heads of institutions, and even lawyers, judges, councilors, etc. "Why all the misery in our otherwise happy home, the Earth," we always ask? Stated, it is the consequence of not obeying the Creator's laws.

What is described above is a tiny fraction of the causes of friction here on earth. Humans think we can do better than the Creator, God. As quoted earlier, "God saw Adam lonely and dejected." He created Eve to be his companion, a comforter, and a soul mate. Adam received Eve. It is reasonable to believe he said, "I'm now complete." You have seen a man or woman who has found the soul mate! Does not the behavior show that God has His hands in it?

A family can't be complete with a father or mother only. Both must be present. But some modern societies want the man and his wife to be co-equals, independent of each other instead of being a comfort to each other. The effect trickles down to the children. Remember, the family is the building block of society, and only healthy families make stable societies. This is a time-tested idea. The level of family disunity and the number of family breakups in 'modern' families bear witness to this. Things are not the way the Maker intended.

The relationship between a man and his wife must be founded on love and respect for God's laws. There are no alternatives. Sure, some men abandon their family responsibilities and engage in actions and behaviors contrary to family and societal norms. Because of these, laws have been put in place that diminish the value and authority of the man as the head of his family. Unfortunately, the effects extend to all family members, particularly vulnerable children.

Some lawyers and family members have exploited these laws because of greed and revenge. Consequently, you often see two opposing camps in the same family: the man and some children unite as one group and the woman with some of the children in the opposite camp, each vying for supremacy. The future earth's caretakers cannot be raised in such families.

Any fair-minded person would acknowledge the love that many men have for their wives and children. The wife and children are the pride of the man.

There is hardly anything closer to his heart! What men have done for the love of women! Honestly, many women do appreciate them. How happy many women feel when men open the door, any door for them, or take them out to dinner or pay the restaurant bills.

Even some women go as far as providing the money but let the men pay the restaurant bills because that makes them feel like women, happy. A woman would not like to sit by herself in a public place, like bars or restaurants. She feels more comfortable with her man by her side.

In the face of danger, men will go to any length to safeguard women and children first before thinking of their safety. How safe a young woman feels when walking by the side of her brother or man in the streets? The reason for this is apparent. Please think of how many songs men have written in praise of women; how many wars have been waged by men to protect their women and children. These behaviors of men are not earthly but from the Creator.

It is common, throughout nature, in the air, land, or sea for male animals to protect females and children. God is Omniscience. He does not diminish but enhances the status and importance of women in the lives of men and the children in their families. He created women in the first place for a unique role, not the role assigned to men. Unfortunately, in some families, mainly 'modern' families, some women see this natural order as an encroachment on their independence. Some women are sold on an idea of freedom that releases them from those "old-fashioned" natural roles. Some even feel disrespected. The result is the never-ending unhappiness in many families.

This feeling of inadequacy or that their never-ending work goes unrecognized stems from some women seeing themselves as disadvantaged and cheated by nature for making them females. Thus, she begins the quest to prove herself as necessary as or equal to the husband or any man. And the struggle for supremacy and the never-ending strife in the family have begun.

But some women feel so demeaned by their role that they even resent being called "Mrs." or called 'women.' You can see why it would be difficult for a relationship between a man and a woman to succeed in such a household. Such families would experience two people living together, with each competing for dominance.

Indeed, we got it wrong, right from the beginning, who the Creator is and what He expects of us.

Let us take some time out to go back to Nature and check out the functions of some living things. Fruit trees produce fruits, seeds, food, and medicines; vegetables have the vitamins, food, and medicines; or trees that provide different needs like shelter, furniture, and other services or animals like the cows, goats, chickens, and fishes that provide food and other services.

These things do not complain when providing the food and services as required by the Creator. They even give up their lives doing their duties. A plant will give up its life when uprooted to provide vegetables; a cow will give up its life to provide meat. When a tree branch is cut out or the whole tree cut down for whatever services, the tree does not cry out, refuse, or resists. It willingly gives up its life. Just imagine the consequences of these things were to reject or resist giving up their lives and fight back. No! They obey the will of the Creator.

Many argue that these things submit to their fates simply because they have no wills or feelings. That can't be true. Watch the reaction of an animal, making a never-return journey to a butcher's house.

The above-discussed issues call to mind the attitude and feelings of humans towards their duties here on earth: The males generally feel or complain that their Creator-assigned responsibilities are overwhelming. Most females believe their gender is their worst enemy. Such women would neither be happy with the 'overwhelming' tasks of men if assignments were reversed. And I don't know any man who can have the patience of a woman or the ability to bear the pains of childbirth and still love the child.

It is the dissatisfaction with our roles that have caused all the divisions in the families, families designed to nurture the future generations of the earth's caretakers. The family is supposed to be the building block of society, a place for comfort, trust, and a home ready to provide support and happiness.

A person without a family is like an isolated person. Without a family, how would one know where he is coming from or going to? The examples cited above explain the importance of a family. It is clear why the All-Knowing God created that unit called family.

These rebellious attitudes show that humans were wrongly informed or did not fully understand who the Creator really is – the Omnipotence and the Omniscience, alone by Himself needing no help or partner.

Besides, He designed humans and, as such, understands all the workings and needs of humans more than they do.

Also, some women object to the assignment of the headship of the family to the husband. These, in extreme cases, have led to the institution of equal partnership in the family, with each operating independently.

Could this arrangement be of the Creator? No! Because in all nature, males protect what belongs to them and assume the headship. This is true with animals that live in the air, land, and sea. The case of humans can't be different.

It is somewhat confusing to hear some women complaining of being powerless in their families. It is like a person inside a river complaining the body does not get wet but dry. What can make a woman, a "mother," feel that way? One plausible answer is a lack of knowledge of the role of a mother in her family as ordained by the Creator. Another is that the woman may be in an unsupported community where the position is minimalized. The level of her role in and commitment to the family will always determine how much influence and power she has in the family and on the family members.

Also, a misguided mindset can make a woman believe that her role, influence in the family to be secondary and unimportant. The inevitable result is waging war for supremacy, independence, or engaging in different behaviors inimical to raising a happy family.

How can a woman realize the actual power of being a woman and mother and take advantage of them in this posture? The role of a mother, any mother, has been explored exhaustively elsewhere in this book. Also, I discussed what men do for the sake of women – writing songs, going into wars, and in times of danger, men can take bullets for their wives and children.

Women are closer to the children of the family, being the person who caters to hunger, the first teacher, moral instructor, and the person who provides the shoulder on which both the children and the husband can cry in times of stress. Of course, every person knows the wonders of a soothing touch or hug of a wife and mother! Does such a woman have no powers, influence, and her role unimportant in a family setup? She has a lot of power and influence!

I don't know of a situation where a woman's (particularly a mother's) contributions or presence is disregarded or belittled in a family setup. How do people generally feel when they see a pregnant woman? In some cultures, like Nigerian, such a woman is almost worshipped. Women, particularly mothers, have the ultimate respect. When a woman drives up a checkpoint in Nigeria, the guards are on their best behavior. Why? They don't know who the woman may be or, most importantly, the statue of the husband. You don't want to get into any trouble with a woman.

In Western cultures, the 'modern' woman abandons her role as a woman and prefers the man's. "What a man can do, a woman can do even better," they claim. In some of these cultures, because of such women's attitudes, their men go elsewhere to look for 'wives,' 'mothers'!

Consider this incident – a personal experience: My wife and I were preparing to attend a function. Usually, I was ready first and got into the car, and waited for my wife. By the time she put on her make-up, turned off the lights, and locked the doors, which I usually forgot to do, and then get into the car, I was already fuming. "What kept you? And did you forget we were late?" I shouted, frustrated. "I forgot to bring your Boost drink. Remember, we don't know when the meeting will end. And being diabetic, your sugar may run low. To do all those was why I had to go back to the house", my caring wife explained.

Please note that the drink was for me, not her. I forgot she remembered; she cared. I looked at that 'angel' with all love and admiration but felt sorry for and ashamed of myself. Is it possible a woman with a similar character as my wife will not be loved, her influence not felt in every corner of the family? The role of a husband and wife is a divine assignment. It is not human! In many households, generally 'modern,' you would have heard expressions like, "If he needed the boost drink, he should have remembered and brought it for himself."

I can't overemphasize that the Creator is Ever knowing. Every individual, anything He created, is unique and designed for a reason. Humans cannot know better than the All-Mighty. We must go back to Nature and read and understand His intentions and obey His laws to end all the unhappiness and misery on this earth.

Considering temperament, judgment, foresight, patience, and the humanness of humans, I had always wondered who should have been the head of families or even governments – the 'man' or 'woman'? Besides, we still hear of "mother earth/nature," not "father earth/ nature!"

The answer can only come from examining nature or God's creations since we cannot verify this directly from Him. A good start would be to find out the arrangements with other living things like animals that fly, live in water, or on land? Almost all point to males to be the head of their families. Why would it be different from humans?

Also, the males provide the 'seeds,' a good reason why the offspring, including the family, take the father's name. In some cultures, men feel complete only if they also have male offspring. The Creator fits out human and animal males with strength, size, etc. Naturally, the father is supposed to be the provider, ensures stability, security, and order to the family.

Some people question why providing the seeds by the males should be more important than the environment that nurtured the seeds? It is clear that without a nurturing environment, the seeds would not survive. There is another way to ask the same question, "Which is more critical the seed that grew into a tree that produced the fruits that fed and nourished us or the fertile soil that nurtured the seed which helped it grow into a tree, enabling it to have the fruits? Without fertile soil, the seed will not grow into a tree in the first place. The answer? Both combined made the seed what it became, a fruitful tree.

The seed and the fertile soil will have to be considered together. Because the two conditions are not mutually exclusive. Also, think of the duties of a 'mother' in the family structure as enumerated and discussed above. Of course, the Creator had known the already existing myriads of responsibilities wearing her down and saw no room to assign to her the headship of the family.

Does this last consideration not add to why the female is never the head of the family in nature? But her roles exert a lot of influence in the lives of the members of the family. Still, she is not the head of the family. The parts played by the father and mother are clear – complementary, as the functions of the seed and fertile soil. They are different but equally important. As the saying goes, 'One should not envy the other.'

There are instances where the wife has better abilities in the management of the family than the husband. That situation does not diminish the statue of the man – he is still viewed as the pride and ostensibly the provider and protector of the family. If the man is belittled, so also will the whole family. Wise and God-fearing mothers know how to walk the fine line and rule 'from behind.'

Chapter 11

NATURE OR NURTURE

It is essential to discuss the roles played by Nature and Nurture, the two critical determinants of the 'future person' a child would become. They also help us have some understanding of what is causing the problems of this earth. We know that before God created the humans, the earth had been made ready to receive them, as legend has it.

Does this statement imply some involvement of time? No, as explained earlier in this book. The earthly phenomenon, time does not bound the Creator. It is more reasonable to believe that the earth and the universe came into existence as the Creator willed them and in the absence of time.

The earth contained all that future living things would need. How can we be sure of this? The Creator of all things is God, the Omniscience and Omnipotence, the All-knowing. And as such, you can't forget or make a mistake. Besides, the Creator executed His awesome designs meticulously. It is evident that, in human terms, He showered a lot of love on his creations. How can any evil come from such a perfect Creator? Then we must look elsewhere for the sources of the atrocities that delude the earth like selfishness, greed, poverty, covetousness, wars, sufferings, unhappiness, etc.

Before I forget, the following need to be examined to understand which they belong: Nature or Nurture. At the beginning of their lives here on earth, all children – red, yellow, white, or black – behaved the same way – like children. Then watch the influence or effect of a traumatic event(s) – like the sudden death of a loved one, flood or fire that destroyed all the child's valued belongings; a rape that left an indelible mark; forced immigration, etc.—that would alter the trajectory of the child's life. These events can be shown to be a direct result of human activities not necessarily induced by nature.

We must examine the effects of nurture in more detail. All living things except humans obey the Creator's laws precisely. This disobedience is because only humans received the gift of choice or will. Was it the ability to choose that ushered in the discontentment, selfishness, unhappiness, wickedness, sickness, and suffering on this earth? Or are these unacceptable conditions on earth the results of disobedience to the Maker's laws? The two questions are the two sides of the same coin.

Briefly, discontentment starts from the home, the smallest but the basic unit of the society, specifically from a dissatisfied partner. Let's take a slow journey into the distribution of chores in a typical family: The father bears the responsibility for the success or failure of a family. A challenging job, indeed! But it is the mother that will endure the inconveniences of a nine-month pregnancy, endure the pains of the necessary natural childbirth, and the subsequent child-rearing, including the essential breastfeeding. In addition to these, she may have to deal with the upkeep of the family and home.

But is motherhood or fatherhood a duty imposed by the Creator, the couple, or self? If willingly accepted, it is the Creator's imposition and will reduce or eliminate the tensions emanating from them.

Males are not designed to carry a pregnancy. Also, they are not equipped to supply the Creator's special milk for the babies. Males, generally speaking, are bigger and stronger and, therefore, must deal with other duties requiring the application of their unique gifts. Then who has it easier in a typical household setup? If men, why do they generally have a shorter lifespan than their wives?

As noted earlier in this book, two warring parties will exist in a toxic family setup, one headed by the father and the other, the mother! Think of the effects of the tension on the children, the leaders of tomorrow, the future caretakers of the earth? They will be molded and nurtured by insecurities, jealousy, unfaithfulness, selfishness, wickedness, and all sorts of evils existing in their environments. It is obvious; these evils will exist in such a family.

Yet we wonder why all the evils and unkindness here on this earth that the Maker crafted meticulously and furnished with all that living things needed and called it the proper home of all living things. Indeed, we got it wrong right from the beginning, the knowledge of the Creator, and the purpose for which He created the humans.

Some of the products of this type of family are our neighbors, homemakers, caretakers, husbands, wives, school children, teachers, principals, professors. They are also the road workers, drivers, clerks, peoples on our roads driving. You can find them in our marketplaces, banks, courts, churches; people that work for or are heads of government institutions; lawmakers, governors, presidents or leaders of countries; private business owners; and the list goes on.

We should not lose sight of families that walk, 'In the way of the Lord' and the products of such families. The people from such families are the hopes of the future generation. They understand the power and knowledge of who created them and for what purpose.

In addition to the influences on families are the media and other effects outside the home. These impact the impressionable minds of children like some of the behaviors which children pick up from other children in the classrooms and playgrounds. Consider Adolph Hitler and his likes – the type of influences that molded and shaped their horrible minds as they were growing up.

On the other hand, think of the type of socialization that produced minds like Michael Angelo, Amadeus, and their likes. These people saw good in art and music. In this group, you can see the likes of Jesus Christ, Nelson Mandela, Gandhi, Mother Teresa, Dr. Martin Luther King Jr., President Jimmy Carter. These people sacrificed their lives or are giving their lives in the service of others; people who believed that justice denied to one person is justice denied to everybody.

These are the effects of nurture. Revisit the chaotic families, those devoid of love, human kindness, human decency, and consider the fate of the children of the families, the next generation of the earth's caretakers. Please think of the foul words that filled their ears daily, as their lungs filled with second-hand smoking and veins with alcohol and drugs!

The children are always lonely, deprived, and ever searching for an escape.

Think of the whispering these unfortunate children will hear daily, particularly in a divorce situation? For revenge, some parents will attempt to destroy or erase the memory of the other parent from the minds of the children. Guess what happens?

Parental Alienation Syndrome (PAD)! We know the minds of children, particularly the very young ones, soak information just like a sponge in water as they listen, day in and day out, to the hurtful words from a bitter parent.

The children are forced to take sides between the father and the mother— a very unwieldy burden on a young mind. Persuading the children to prefer one parent against the other will be equivalent to dividing the children's minds mathematically into two. The child bears the blood of or is a product of the two parents. How can the child choose one parent and despise the other?

What type of men and women would children brought up in such a divisive and hostile environment become? Unfortunately, some parents don't care. They must vent their anger and get their pound of flesh!

Those parents did not understand or care for what marriage was about – an institution by the Creator Himself – neither did they know who the Creator is. Would these be good excuses for destroying innocent minds? No! All of us have the same gifts of brains, senses, and conscience. To use them is our choice.

A child's mind can also be poisoned against other children by the indoctrination from the parent(s). "Johnny, you can't play with Love Young—they are bad people, or we are better than them." "But Mommy, he is my friend," the young mind-objects. This is worsened when no objectionable behavior of the forbidden friend is described but only skin color or social/economic classification.

We know that children are like Angels, servants of God. No evil is in their minds, just like a clean slate without writing on it. Then slowly, their impressionable minds are filled with garbage – hatred, pride, etc., from the parent's teachings (s). Hear anything repeated many times, and you will begin to believe it to be true. So, the future earth's caretakers will begin to internalize all the garbage and soon identify with them.

As a result, one day, we will hear that a young man or woman picked up a gun and started shooting indiscriminately at people he never even knew. It could be he turned the gun on his parents or even on himself. Why? He saw everybody as bad people. Children from this type of dysfunctional family will be involved in cut-throat deals, I-must-win-by-all-means businesses where winning is everything, and there is no room for failure. They must be wealthy and remain so because the poverty and the suffering of childhood days must be banished forever.

People brought up in these conditions will have a heart of stone to move into other peoples' lands, homes and destroy, kill both children and women, people they never met or knew, and occupy their lands and take over all that belonged to them. They rule with iron hands in their loot, rape, and killing spree on the high seas.

They can declare and lead men to wars on flimsy excuses because they have 'no' moral values and are bloodthirsty. Their consciences were devoid of human decency; they have no qualms, no respect for human laws or the Creator's. You would not even call them brute wild animals – wild animals don't go on a killing spree or try to usurp others' territories. Why? All animals but human beings obey the Creator's commandments.

Other children are raised to believe they are the center of the earth – others must worship them. They are too good. Just imagine what unhappiness their attitudes will bring to people around them. You may have encountered such individuals in your everyday interactions with people.

Consider a child brought up under high moral ethics – to know that this world belongs to all of us. Then he would believe all the Creator's bounties belonged to all of us and should be shared equitably among all living things and for all generations. Such a child will grow up to be a happy and considerate person. And the world around her will feel good and comfortable.

From these examples, nurture or the environment in which a child was raised plays the central role in determining who they will grow to be. Thus, by extension, the good or evil that will emanate from him.

Please note here that some willful children go astray despite nurturing in a good, stable family by choosing to be nurtured by negative outside influences. The opposite can also occur, i.e., children from non-nurturing families can willfully seek positive nurturing situations. Yet and still, there are predators who, under the guise of nurturing, try to corrupt innocent children. It is not Nature. Nurture is undoubtedly the culprit, the primary source of all the evils and unhappiness we experience here on earth.

All human beings were created good by God because the Creator, God, in human terms, is good. Humans, by their nature, are good! This is the only conclusion one can reach after examining God's creations around him: how much He cares for His creatures, His abundant and ever-lasting provisions, and the beauty displayed in His creations. God is ever knowing. He knew all His creatures needed to keep them happy and satisfied.

He did provide all that living things need and the means for acquiring and using them. He left nothing to chance. Above all, He provided everything free of charge. All He asked of His creatures was to live here on earth, forever and happy, and to take care of all His creations. He is indeed a kind and loving God. How can the evils and unhappiness we experience here on earth come from Him, Nature?

No! It is from lack of knowledge of Him and disobedience to His laws. We were not taught right from the beginning about the Creator by our religious teachers. However, we are equipped with the intelligence and the senses to discern and make our choices.

At this point, we must examine another source of discontentment in this world brought about by how humans process or accept some natural phenomena that are called 'extremes of nature or outliers in creation. In short, these are called 'Nature.' Nature is the set of influences imposed by the Creator, which humans have no control over like: being born black or white; short or tall; Japanese or Iranian; gay; bisexual or transgender; etc.

The Outliers or natural extremes have been misunderstood since humans appeared on earth. Ignorance is to blame, but time will solve this. There was a time when twins were rejected as bad luck or evil omen. Their people killed them, and their mother suffered the same fate. The same treatment was meted out to lesbians, gays, transgender, little people, giants, albinos, and extraordinarily dark or white peoples. These differences are treated under outliers and infinite variety elsewhere in this book. There is no problem with Nature! Our level of understanding of 'infinite variety' in creation is the problem.

As noted earlier, no two things are the same. God did not create anything that way. Only the Creator is one. It is clear then that we are not born equal. Did the Creator create kind people, wicked people, loving people, etc.? Or put differently:

Is it possible that God created some people good, some kind, some with despicable behaviors, etc.? If God created such a person, how can the person be held responsible for his actions? These questions have troubled many minds. Please consider the following to answer those questions.

God gave humans the ability to develop free will, with which they can make choices as their heart desire. The Creator does not abandon a person because of the gift of free will. No! He also gives the person the ability to develop the senses, the intellect, the conscience, etc., to guide the person.

Two 'different' people to become 'one,' raise a family, the earth's future caretakers would require obedience to the Creator's laws, love between the couple, unity of purpose, and determination to do the right thing by the family. As said before, this is not an easy task. But it is achievable, and many have done it successfully.

We learned that it is the law of the Creator for all who can to marry and raise families to keep the Perpetual Cycle running forever. Through studying Nature around us, we also learned that the Creator gave authority to humans to control and regulate the welfare of all living things. However, each family will have to decide on how best to run their family. It is an obligation to raise the children 'in the fear and respect of the Creator. Above all, love must reign in the family. These laws are of the Creator.

Finally, it must be pointed out that the children – future earth's caretakers – also have an obligation in this. The Creator gave the same gifts of the brains, senses, the conscience, etc., resources required to guide humans in their choices and decision making. Therefore, all decisions they make carry the same rewards and consequences. Everything hinges on knowing who the Creator really is.

Are we of the same image and likeness of the Creator? If the answer is yes, our behaviors or how we conduct our affairs here on earth will reflect what was learned from our environment (nature), the same way a child mirrors what he learned from the parents? It would not be misleading to say that the 'parent' of humans is the Almighty God, the Creator. We cannot see or talk to the Creator as a child can to their parents, unfortunately.

The communications between a parent and the child are necessary because there is no other way a parent being human, will understand what is going on with or teach, guide, and protect the child. The relationship between humans and God is, of course, totally different. God is a spirit and the designer of human beings. As such, he does not need to talk to or be seen before communication with His creatures can occur.

Does not the designer of the car understand his invention? God can read the minds and hearts of individuals or however He pleases. Doing this is even not necessary – God is the Omnipotent and Omniscient. So, before God created crated a person, He had known all about that person.

We know that God designed everything that exists and established his laws, including setting up rewards and consequences. He also instituted what I would call the Ns factor to guide all living things and provided all that is necessary for all living things to live and accomplish all their tasks here on earth successfully. This arrangement, in part, shows God loves His creations.

For a successful education of a child, part of the instruction must be practical examples.

Of course, God gave humans the characteristics of learning this way through senses. The Creator provided examples for humans to follow.

Our fore-parents saw a seed germinate in the wild, grew, and began producing the fruits or crops they ate previously. They repeated the process. Observing and repeating this is how farming was born. Humans only looked at an existing example, the eyes, and copied from it to design the camera lens. What of developing the computer, they copied the workings of the brain.

There is another excellent example: To build a house or any structure, they copied from their 'parent,' the Creator. How? God first 'built' the skeletal system and attached flesh to the skeleton and covered and beautified the whole structure with the skin. This description is the only way a human being, an earthly creature, can explain this miracle. God does not build. He creates. 'Observing what the Creator did, humans learned that to build a house or any structure; you need framing (the skeletal system) first, then cover with drywall and paint (the skin).

Fig. 8: A Collage of Different Types of Flowers

These beautiful flowers are a few natives of my home State. Check out the flowers in your neighborhood.

I'm sure they are as beautiful but certainly different types. Just imagine how many kinds of flowers in existence on this earth. Again, the number approximates infinity.

But why these different types of flowers with other characteristics? Flowers can help humans have a peep into the intentions of the Creator - His care and love for His creations. The beauty of flowers comes from the color, smell, and functions: When one looks at or smells a flower, all the senses are activated, and the heart gladdens with joy. One derives the same experience from some grasses, leaves, and some barks of trees. From some of these plants, one can extract aromatic scents. The intentions of the loving Creator are apparent – to provide joy, happiness, and satisfaction to all His creatures living here – 'Paradise Earth.'

We see the amount of love and care parents (including animals) give to their children – providing all the children need for their wellbeing. Where did parents learn how to love their children? From the Creator as etched in their DNA. No parent, no dog or ant, etc., needs a college degree to know how to raise her child(ren).

Every parent knows to: provide food including medicines, clothing, protection, materials to build their shelters, teach them how to procure those for themselves when they are on their own, protect themselves from predators, teach them the importance of procreation to keep the presence of their kinds here on earth forever. The parent teaches as she learned from the Creator.

It is, however, more accurate to say that parents only direct guide their children to these behaviors. Remember, every child possesses the ability to develop these behaviors as already in the DNA. And they grow and realize them as they gain maturity and experience. They need their parents' help to learn them.

Who taught a cat, a goat, a cow, an elephant, etc., how, and when to get pregnant, give birth, and raise its young? Did they read some manuals? These things are written in their DNA by the Ever Knowing.

He did not merely create living things and their needs, assigned their duties, and left them on their own. We know this from his provisions, including the intellect (intuition), senses, conscience, etc., to all living things and because no human parent can do the reverse to their children. Then how can the Creator, from whom they learned, do so.

Considering these provisions, care and love by the Creator, I have always wondered from whom humans learned certain behaviors that have been the root causes of human suffering: Behaviors like selfishness; cheating; little or no love for all living things, particularly fellow human beings; deriving pleasure from the distress of one another and even from the sufferings of helpless animals?

From whom did the humans learn to cheat or give in half measures? Of course, not from their 'parent,' the Almighty Creator! As identified earlier, these behaviors humans learned from the events in and outside the home. We have come to know that any human practice outside of the Creator's design incurs dire consequences— the sources of the discontentment and unhappiness we experience on earth. Yes, the

Creator 'loved' His creations so much he provided all they needed to live here on earth permanently: shelter, food to eat, and a means of self-protection.

The Creator placed humans in locations so vast that one can live anywhere in his assigned environment and outside of it temporarily. Check out the distribution of all the life-sustaining resources: food, water, air, sunlight, and soil. He provided for all creatures. Above all, He provided humans with a great gift, intelligence. Indeed, humans did not learn their destructive behaviors from God. We must use our knowledge to understand the Maker's intentions regarding our relationship with one another, mainly how to distribute or share the earth's resources amicably.

Consider the gift of water; for example, The Designer did not allocate 2% to one family and 60% to another or 40% to rich people and 1% to poor people. The quantity of water is such that every individual, every organism, has access to as much as necessary. Humans don't share in that manner.

Also, for Sunshine, nobody can claim ownership of or allocate a certain percentage to himself. This is true with all the gifts of creation. The quantity, the color, taste, etc., of mango fruit, pear, orange, etc., is not determined by who owns the tree. A tree does not determine the number of fruits it bears on whether the owner is a black, white, yellow, or red person. The fruit trees are what they are.

What should human beings have learned from the Designer? That His creatures should share in His abundance equitably. Only humans don't imitate the Creator in the distribution and sharing of the Creator's abundance. Take, for example, the lions: They don't go about in a killing spree, killing any animal they see since they have the power and agility to do so. Any lion that behaves that way outside the norm is not normal, and the earth's caretakers are empowered to get rid of it. When hungry, they take just enough. No lion or any animal, in normal circumstances, will contravene this law of Nature.

If humans are in the image and likeness of the Maker, fairness, equity, and above all, love will be the hallmark of all their actions. Also, if humans are in the image and likeness of the Maker, they would be spirits following His examples and obeying his laws like the 'angels.' Then the earth would be a happy place as the Maker intended.

Indeed, we were taught wrongly who the Creator was, right from the beginning!

One thing is sure that the earth will eventually be the Paradise it was created to be. And it is assumed that Science and Technology will catapult our world to that state. One would ask how that could be possible considering the level of non-compliance with the Creator's laws - the advances in developing the instruments of war and the persistent attempts by humans to alter the Creator's designs?

Even some fear the earth, following this trend, is destined for self-annihilation, the ultimate consequence of the advances in technology. Considering these uncertainties, how can we be sure that the earth will eventually become the Utopia where suffering and any form of unhappiness will be absent and only everlasting bliss?

First, God is God, and nothing can change or alter His designs or intentions. He has divine deterrents in place to prevent that from ever happening. Because we live for the moment, we don't take cognizance of why He bounded all living things together with time. Everything will require time to reach fruition. Everything! So, fulfilling the Creator's intentions—the Paradise earth—will take time to come to fruition. It cannot manifest itself suddenly. Let's review some phenomena.

Peoples have finally started to comprehend the Creator's intentions. A few examples will suffice to make this point: Think of people who spend all their lives amassing wealth at all costs, but they decide to give it all away towards the end of their lives. Why? All the wealth will mean nothing in the end, they had realized! What he took as his did not belong to him but to the Creator who had provided it for the sole purpose of aiding living things to live and perform their duties here on earth. Nothing is for hoarding by anybody. It's God's law that no living thing can take anything with them when they die.

When a person dies, all the cars, mansions, planes, gold, diamonds, and silver will be left behind. Humans have examples of the ancient Pharaohs, but still, they persist. The question? Why would any sane person spend all their lifetime cheating, killing, not enjoying himself, etc., to acquire what would mean nothing at the end of it all? Still, we wonder why all the suffering and all the unhappiness.

There is nothing wrong with amassing wealth or inheriting it. There is nothing wrong, being rich. The Creator, from whose examples we should learn to guide our behaviors, has everything. In human terms, He is the wealthiest. Therefore, there is nothing wrong with being wealthy. Possessions of any person (poor or rich) belong to the Creator and benefit all His creations.

Check out the following: occasionally, we witness some acts of love, mercy, selflessness, sacrifice, etc. The media and private citizens hold doers of these acts as heroes. But, doing these always, not randomly, are commandments from the Creator. We know that failure carries consequences (and humans fail always)—the emptiness we experience here on earth. Considering the enormity of these failures, it is surprising that many people still wonder why all the sufferings we witness every day in many places. The simple answer is that humans continue to violate His commandment to "Take care of my Earth."

It is easy to believe, though erroneously that the Creator guides humans in all they do. No! Because that would imply that the Creator guides peoples to commit evil or good. Put differently, some people believe that the Creator guides every good deed but an evil act by the "Devil."

In this context, where is the conscience, free will, intellect, senses, etc., given to every human being by the Creator to enable them to manage their affairs here on earth? What of the Divine deterrents? All these checks and balances are already built into the system. Then it is reasonable to deduce why the Creator, God, did not need to send angel police officers to live among humans, monitor their everyday behaviors and activities, and dispense justice.

Besides, God did not build his system to interfere with the free will He bestowed on humans. So, it is unreasonable to believe that our Creator guides people to do anything good or bad. But all the resources one needs to do anything they choose to do are provided. We make a choice; then experience the rewards and consequences. Being the Ever-Knowing, he did not just confer free will and leave humans to fend for themselves. No! He provided all the checks and balances so that humans will discern what is right or wrong and finally make choices. This arrangement shows that the Creator 'loves' His creatures by providing all the tools they need to succeed here on earth. So, the choices we make are our own – not guided by the Creator or induced by the 'Devil.' Our options are only limited by our inability to control nature's rewards and consequences for our actions.

Though humans have free will, they can't exist even for a moment outside the watchful 'eyes' of the Almighty on His creations. How do we know this to be true? What can living things do on their own outside of God's realm? Nothing! Humans cannot create crops and make them yield food. They can't digest the food and extract nutrients and send them to the cells where they are needed, on their own. Humans or living things can't create the sun, air, water, etc. Humans can't even manage the free will they are given. A good analogy for this Creator-Human relationship: A parent teaching his child how to ride a bicycle. When the child is placed on the bike, the parent holds the bicycle, guides the child, and slowly releases his hands while still running behind the child with his hands instinctively stretched out. You get the point. Though we are given free will, senses, and intelligence, the watchful eye of the Creator is ever-present.

The undoing of many humans is their inability to take responsibility for their destructive behaviors. They love to shift blames to another person; they prefer someone else to suffer the consequences of their actions. Some people go a notch higher by believing that they, being the children of a merciful God, can commit any evil, and the Creator will look the other way. The Supreme Lord of Justice will look the other way! That's impossible. Therefore, we hear, 'the Devil made me do this or that.' Or all I need to do is confess whatever sin I committed and continue sinning. There is nothing in nature to show that the Creator so desires these or that He will look the other way.

Unfortunately, all this nonsense does not relieve humans of the consequences for defying the Creator's laws – the unhappiness seen in many places. We know the Almighty God can't be deceived. And we know that those who comply with the laws of nature have nothing to worry about and do sleep at night with their two eyes closed.

There are many such peoples.

One familiar story, an act of kindness and selflessness, the very expectation of the Creator goes like this: A woman attended a function held in a large football stadium. When she drove into the stadium park, she realized she was the first attendee. She packed her car far away from the entry gate. During the twenty-minute break, a friend saw her walking down to her car. The friend was confused about why a person who arrived first should park so far away from the gate.

The friend could not help himself and decided to ask her why. "I thought you told me you were the first person to arrive here today." She replied, "Yes." The surprised friend asked, "But when you arrived, the lots near the gate were empty?" thinking she was nuts for not taking the advantage to pack near the gate when she was the first person and the lots near the entrance empty. The man was almost shocked when the woman told him her reason for packing so far away. "My friend, remember, some people will arrive late and therefore in a big hurry. They would be the people who would need the lots nearest to the gate."

The man could not believe what he just heard. The behavior is an example of selflessness, caring for a neighbor you never met, and caring for living things as God commanded.

Humans are the only living things allowed to develop the intellect and the senses for a purpose—to take care of living things as they take care of themselves. Do we take care of others as we take care of ourselves? Only sometimes! What of respecting all living things? Again, NO! But other animals obey God's commandments. Hence humans cause their misery on earth—consequences of not complying with the Creator's orders.

One more example: One day, a group of men on the road construction assignment confronted an unusual event. As they were clearing the bush and taking their measurements, they got to a point where they were surprised by the presence of a defiant bird that stood in front of them and audaciously refused to fly away and make way for them to continue with their work.

Upon investigation, they discovered the bird had laid eggs under one of the bushes. The men did not drive it away, the expected behavior of some humans to such small, inconsequential animals. Some people, confronted with this kind of challenge, will get rid of the 'little insignificant' animal, paying no heed to the commandment, "Thou shall take care of my earth."

What did the caring, nature-loving road workers do? They stopped construction close to the nest and eggs, built a fence around it, and continued their work from the other side. A few weeks later, they found out that the eggs had hatched, and the mother flew away with the young ones. Then they completed the work in the area occupied by the nest. That was the love and care as required by the Creator.

Have you witnessed a lion playing with the cub or a human mother or father playing with their young; have you seen two lovers playing together? What of the 'sea monster'— the shark—playing with the young?

Perhaps you have witnessed a giant crocodile transporting the just-hatched tiny young ones on its razor-sharp teeth from the sand where it was hatched to the water? An amazing spectacle, indeed! With its giant teeth, a crocodile could tear an animal to pieces for food or defense. But its young can ride safely and comfortably in its mouth.

Is there any other conclusion than that the Creator desires that love and caring for all living things be the hallmark of human behavior here on earth? Unfortunately, humans failed to copy 'Mother' Nature, the Creator. Doing the Will of the Maker is not a matter of "if I like." It does not mean that "my God is kind and will forgive any sin," so I don't have to do the Creator's wishes. No! It's a matter of being a part of the natural order. But a disrupter, the disobedient, will suffer the consequences in the form of the misery we see here on earth.

These are examples of how it should be – to love all created things as we love ourselves. There should be no 'them and us.' Nothing in our world can show the Creator desires that we should do good only to those we expect good in return. Humans certainly got it wrong, why God put us here and what is expected of us.

It is hoped that a time is coming when humans will realize that the purpose for which they were created was to take care of themselves, one another, and the earth. Then, powerful nations and individuals will no longer compete in amassing instruments of destruction, acquiring wealth by all means, careless of who gets hurt. Then they will compete in 'goodness,' and who will be first to send help to a more impoverished nation so that they can get out of poverty. They will find cures to eliminate all types of diseases plaguing the earth instead of exploiting sickness for excessive profits.

Then, many rich countries will include the welfare of poorer countries in their annual budgets, with fair trade for their resources and labor. Individuals will not forget their neighbors in all they do or in their thinking. Then the Paradise Earth will finally manifest itself.

As explained before, examining God's creations – all the living things; all His provisions; the equity shown in the distribution; the everlasting nature of the provisions; how He tied all together in an eternal bond; it is further proof that His intention for doing all these was to have all He created exist here wholly and forever happy.

It is correct, therefore, to assume that the law given to all living things by the Creator was, "I, your Maker, have already provided you your food and given you the means to get it and the means to protect yourselves. You must go for the food, and you must eat it to live; you must continue your kind and fulfill your duties here on earth. And to humans, he added, "You will take care of my earth and protect all that is on and in it. And because of this extra assignment, I give only your kind, the ability to develop the brains, the seat of wisdom, the senses, and conscience, etc., to aid them to accomplish the added task."

When our ecosystem is examined—it is also evident that man is meant to live here on this planet earth and forever where all he needs to survive is provided. The rules/laws to guide our behaviors, the rewards, and the consequences were put in place even before the Creator put living things here on earth. Then comes a moral question: If the Creator provided all we need to live here on earth, happily and forever, why all the suffering, poverty, killings, unhappiness, selflessness?

First, God is God and therefore knows everything and does as He pleases. Second, they are consequences for disobedience of His rules. He knew some humans would obey and receive His rewards while others would not, and they would suffer the consequences.

Some people have questioned whether global events like the earthquake, volcanic actions, floods, etc. are consequences of human indiscretions? Not necessarily so, is the short answer:

1. God is God and knows more than we do and can do anything as He wishes.
2. Some of these 'Natural Disasters' are caused by human activities.

3. Through examining God's creations, we know that the All-Mighty Creator is good, including all He created; His intentions are for all His creatures to live here on earth forever and happy.

Therefore, we must believe that whatever He does must be to fulfill that purpose.

Our beautiful earth can indeed get warmer than as designed by the All-Knowing Creator by the Green House Effect, a consequence of human activity. The warmth can reach a level that can cause Ice to melt, predisposing Oceans, Seas, and Rivers to rise above their banks, low-lying flooding areas of the world. This event will cause death and destruction. Is the general question not: 'Where was God when the death and destruction took place"? God was supposed to have prevented such a disaster from happening! He was supposed to have guided humans or the perpetrators away from their behavior. God does not tell any human what to do.

These expectations are not realistic. God gave humans the brains, senses, and conscience. These gifts were supposed to be used to guide their actions and behaviors. Humans only have a choice in whatever they choose to do. They can or may not induce the Green House Effect, which predisposes to the consequences of the earth warming phenomenon. God does not guide any person to do good or bad.

We were not thought right or mislead into believing that the Creator will guide us to do evil or good? Whichever is the case, the responsibility rests on the perpetrators because they have the tools to make their choices.

The design of our planet is complete, meaning that all the needs of the living things. Therefore, God does not need to send His angels to monitor the day-to-day behavior of all living things.

He created the humans and blessed them with the seat of wisdom, the brain. The Ns factor is a guard that prompts humans to take care of His creations. Humans are endowed with the ability to determine what is right or wrong and empowered to set limits and impose consequences as they see fit.

Humans must protect all lives – plants and animals. All lives are sacred. Because God created them and for a specific purpose; therefore, no life can be taken for granted. If a person/any living thing is a danger to others, such a person or thing should be dealt with by the laws of the land.

Only humans disobey the laws (natural and human-made), particularly laws about our relationship with one another and other living things. All living things should take just enough from the Creator's abundance, no more, no less.

But think about the reality – one person would like to own all the lands, all the wealth, and even want to own other human beings. This behavior is not of the Creator.

If you have come so far reading this book, you know enough to agree that most of the problems of this earth have their roots in Nurture.

Chapter 12

CULTURE

Devotion to or ways we worship God generally describes faith or religion. The practice is done in different ways, depending on the community. Hence it becomes a part of the culture of the community. Culture defines the people of a community. Culture is the unique way an ethnic group does things – the way they worship, their language, mode of dressing, talk, cook, dance, their marriage, birth and burial practices, ceremonies, etc.

Many cultures remain intact unless tweaked, generally for the better, by the peoples themselves. But on examination, the cultures of most of all colonized peoples were affected by the cultures of the colonizers. In Nigeria, for example, European Christianity was firmly established there, particularly in the Southern part. This new religion affected the culture of the natives in no small ways.

I know many people may agree that Christianity is a foreign religion in some African Countries. But the part played by African Christian Patriarchs like Saint Francis of Assisi and North African Bishops in the Council of Nicaea in AD 325 shows it is not so. Yes, some countries in North Africa only like Ethiopia! But in Sub-Saharan Africa and many other African countries, Islam and Christianity are foreign religions.

Even now, people still worship in their traditional (native) religions. For example, all 'white' weddings in many parts of Africa must follow traditional weddings. Or put differently, no marriage is valid without the traditional wedding, which adheres to the native customs and laws.

Before the advent of Christianity, each community followed the precepts of their religions strictly. The current evils: cheating, lying, murder, abortion, embezzlement, family conflicts, divorce, etc., were almost absent because their beliefs forbade such wicked behaviors.

Families and individuals were held together by the devotion to and respect for God, the loving being that provided all they needed. Men and women tried their best to be upright because they dare not displease their generous and loving God. Everything was fine.

I still remember when treaties, agreements were concluded by the shake of hands or a nod of the head. Then a man's word represented him and told of his relationship with the Creator. People strived to be seen that way.

In these cultures, they don't just have one God, the Almighty. No! They have other deities. Chi-ukwu (the greatest or Almighty God) is the same as Chi-neke (God the Creator). There are other deities, and they are called 'Chi' without the 'ukwu.' These are the spirits of past parents, grandparents, etc. They are intercessors. In times of great need or grave circumstances, one called on them to approach the Chi-Ukwu and beg for solutions.

They did so because the people, ignorantly, believed that the Chi-s have direct access to the Mighty God, and they also rely on them for guidance and protection. Of course, all these Chi-s are spirits, they believed. There are other spirits besides the Chi-s. These spirits are direct messengers of Chi-Ukwu only.

My father was not only an herbalist but a 'Medicine' man or a Native Doctor.

Among his instruments of trade were idols or carvings representing most of these Chi-s. He used to explain what each meant when we, the children, became inquisitive.

It is interesting to notice parallels when one visits many other places of worship. The idols, statues, or symbols seen in these other places are of different make and names, of course, but they serve the same purpose. Those tell me that all peoples worshipped the same Creator but in different ways. The different ways they go about achieving the same end are understandable, considering the 'infinite variety' logic.

These people saw the Spirit or Power or Extensions of Chineke (God, the Creator) in everything created like the Sun, Rain, Rainbow, Thunder, River, animals, etc.

Some communities go as far as offering sacrifices to or even worshiping some of these creations.

The communities don't worship Chi-neke or offer their sacrifices to Him. They believed that their sacrifices were not even worthy of the Mighty God. How could an inadequate earthly beings have direct access to God, they wondered? This belief and behavior make sense, considering that God is a spirit, but humans are mortal?

The name, God, was held in awe and hardly mentioned. Therefore, everything was directed to the intercessors, ancestors (deceased relatives), through whom they had no fears addressing their pleas and sacrifices. This behavior did not take away anything from their belief that God Almighty is their Creator, provider, and supreme being. He stands on his own and needs no help from any of His creations.

In Igbo land, this belief is depicted in the names of almost all the children from the community. A child's name will include a reference to the Almighty God, the Creator, and an implicit story attached to it. So, most Igbo names are a combination of the word, God, and some explanation of a prevailing circumstance or a current event when the child was born. Interesting!

The writer of this book was named "Onye-ma-Uche-Chi-ukwu." This five-word name is written and pronounced as one word – 'Onyemauchechukwu.' The compound-word name means: Who (Onye) knows (ma) the will (Uche) (of the Almighty God (Chi-ukwu).

I was born seven months premature after World II. Think of the level of sanitation and availability of adequate nutrition then! I still feel sorry for my parents – how much they must have suffered to keep me alive!

A child born under such circumstances had a 50/50 chance of surviving. The implied meaning of the compound word, Onyemauchechukwu, is, "Who knows the will of the Almighty God about the survival of this child born under those conditions and circumstances. In other words, only the Mighty God knows whether this child will survive or not. The people's belief in God and justice was written in stone, unshakable.

I witnessed one incident in the village: a young man stole some money: the punishment was that he should be taken to and disgraced in the marketplace. The family (the father) was supposed to bring the young man out to the community to receive the punishment. According to their religion (custom), the father was warned that he would directly receive the punishment himself if he failed to bring the young man out. The above-described practices were only glimpses of the level of or how justice was seen and applied in those communities.

Contrast this with what would have happened if the same situation took place in the West: All the young man's relations would have come together to make sure he was not found guilty. Think of how much money would be spent on legal fees to make sure they twist the truth in favor of the young man. People of this culture, Westerners, will see nothing wrong with dispensing justice that way because it is their way of life. But an Igbo man will cringe.

No two cultures are the same. Each is good but good for the community that established it. You see varieties with all human-made laws – they are different in different communities and can change or be modified. But the Creator's laws are the same in all societies, universal. Consider these two examples – eat and sleep. No community does not eat or sleep. Therefore, it is of the Creator. Culture is synonymous with the people of the community: any interference spells disaster.

Many aspects of non-Western cultures changed when big money, lawyers, and religion took over the legal system. Previously, if anybody intentionally took the life of another, the law was that he would pay back with his own life or make reparations or even be forgiven (to live with the natural consequences of his wrongdoing) as the bereaved family demanded. But since the introduction and adoption of western cultures, lawyers, bribery by big money people, foreign religions took over the justice system. Even the impeccable Oracle, the Guardian of the Land, was compromised! Sad!

In most Western Courts, an accused does not have to appear in the court proceedings if there is a legal representation or answer any questions without the approval of their council. In my village, people will wonder why the system was not looking for the truth but colluding with the accused to escape justice.

The answer is simple: You need your lawyer's approval to avoid self-incrimination. In some cases, people can take the "5th". Nobody in Western culture will quarrel or find anything wrong with this method of justice delivery. It is their culture and good for them. You can appreciate the dilemma the God-fearing African natives were saddled with when Europeans appeared imposing their way of life.

Religious laws regulated even the behavior of family members. Children were expected to be obedient to their parents and wives, their husbands. The people who made and enforced these laws included heads of families – the fathers – who were expected to be honest and upright. If they were not, the Oracle, the Guardian, Chief of justice was there to impose their punishment.

There were Order, peace, and happiness. Every person knew their place in society. Also, there were disputes, a by-product of human interactions. What could not be settled in the family was referred to and resolved by the village council. Each community had its oracle that guarded and supervised the committee. And their decisions were supreme.

Contrast that with what happens in the West, where the political parties make every effort for their party to nominate the Supreme Court Justices. The more justices a party appoints, the better for the party. Remember, the Supreme Court has the final say in any case. These Parties believe and hope their nominee will vote in their favor in a contested case. You find this type of justice system in Western cultures.

People in my village lived a communal and straightforward life. Everybody was the keeper of his neighbor, and it was the village that raised the children, a task impossible for one individual. Children were the responsibility of all the adults of the community. If a child steps outside of the communal norms, any adult present was expected to deal with the situation, including taking the appropriate disciplinary measure. If the adult failed to rise to the occasion, the blame goes to the adult. The expectation was a grave matter. Children felt safe, protected if an adult, any adult, was present.

But think of situations where all one needs to do was 'confess' one's sins to someone else and assume the sins forgiven. The peddlers of this belief teach that all humankind are sinners. One can sin as many times as possible. Sinning was natural, and everybody did it. All a person needs to do is confess the sins to get them forgiven.

The new religion's adherents also believed that the Almighty God is kind and merciful and will forgive their sins any time, no matter how many times! Does this belief even make sense? I guess it does but to the people of that culture. It is evident why some people go on a 'sinning' spree, not obeying the Creator's laws of "live and let live." Hence incurring the divine consequences in the form of human suffering and guilt here on earth today.

Is there anything in our world, nature, to prove that the Creator intended to allow humans to disobey Him continuously, confess their sins, and He would forgive the transgressions? Nothing! Think of the effect of this belief on the simple-minded peoples of the Third World countries when this religion, "the only region acceptable to God," was sold to them! Indeed, we got it wrong or were misled by the peddlers of, "We are all sinners and only need to confess the sins and 'continue sinning.'"

I still remember when groups – men or women—took turns to help each other. Generally, the women take turns inviting other women (of the same age group, for example) in her community to help with some tasks – planting, sowing seeds, painting, decorating her house, etc. They lived a communal life, remember.

Check this out: If a nursing mother had to go to a market that took hours, she usually would leave her child with another woman in her neighborhood to take care of, including breastfeeding the child along with her own. Neighborliness was such a common practice then.

A woman was loved and respected most when she was pregnant. During pregnancy, women had the upper hand in their families – whatever the woman said or demanded, the husband had no choice but to obey. When irritated or upset, a pregnant woman could do almost anything. The husband would instead beg her to calm down.

The sight of a pregnant woman then gave joy to all the community members: she was bringing a new life into the community and increasing their number at the same time. These changed with the advent of western cultures – pregnancy became an inconvenience. And being your neighbor's keeper turned into 'self-first.'

As stated earlier, each community of the world developed their own cultures, and they were good for them. There is trouble only when an alien culture is imposed or used to displace the indigenous culture. The imposition was seen in all colonized countries and has remained a problem in these communities.

Remember the word "pagan." Remember who made this pronouncement and who was called a pagan? The term describes a non-Christian, a heathen, a Muslim, or a Jew. Many people from African and third-world countries were called pagans. The people mocked as pagans were the very people who practiced communal life and were 'your brothers' keeper.' They were the people who practiced 'live and let live' and recognized that they were nothing without the Creator. But how could people who did not see the need for usurping other's things; or believed that 'we are all sinners and only need to pray or confess the sins to have the sins forgiven be called 'pagans'? People who believed that all transgressions or rather disobeying the Creator's laws carried consequences should not be labeled pagans.

These God-fearing peoples were called pagans! Indeed, we got it wrong, or our religious leaders misled us. All the native religions were the right religion.

Invaders of these third-world countries did not obey the laws of the Creator. Or else they would not have invaded these places, killed some of the peoples, and plundered what belonged to them. They would have instead learned or copied from the cultures of those obedient to God. Disobedience of the laws of the Creator carries consequences – unhappiness, the breakup of families, etc. Because these simple-minded peoples abandoned their Creator-inspired cultures to accept the invaders' ways, they suffered the consequences.

As we know it today, money was not a common commodity then, and few people cared for or needed it. Really, what was the use for the money? I used to think of the statement, "In those poor countries, people lived on less than two dollars a month." In some cases, you could put the number to zero. The statement itself shows ignorance of the mode of living or lifestyle of the people. Why?

Almost all the people needed, they produced themselves from their farms and gardens. Each family had its farmlands and gardens. In these places, they planted all the required food items: from vegetables of all types to peppers, seeds like beans, groundnuts (groundnut oil), corn (corn oil), garden eggs, tubers like the yams, coco-yams. They also planted bananas, plantains; palm trees (palm oil), pear trees, cola nut trees, oranges, mangos. Even animals like chickens, including eggs, goats, dogs, cats, etc. they raised in their farms. Other types of animals were not raised by families like 'bush meat' (the grass cutters), deer, and snakes. Depending on the locality – cows, fish, rice, etc., were produced. Why did anybody need money?

The farm products were traded (bartered) for other items that a family might need, like tools, textiles, and building supplies. My father sent his first son to college from the money he made from the sales of his agricultural products.

Some of the food items were seasonal except for some vegetables like the Ugu and Uha, including some animals like chickens and goats. Food was plentiful, particularly during the months of harvest. Food was in short supply only after planting seasons.

People knew how to prepare for such seasons. For example, people did not plant all sizes of yams or coco-yams. Small sizes were generally suitable for planting and larger sizes for eating. So, most of the big ones they preserved for use during and after the planting seasons. These, combined with vegetables, including seasonal ones and fruits, seeds, cassava, which they can consumed in different forms, meats, etc., will still be available.

Despite these measures, some families were still in poverty. I'm not among those who believe that the so-called destiny determines family situations. Why? Because God provided equitably to His creatures to be able to live here on earth happily forever and be able to complete their duties.

When I was growing up, there was plenty. After yam harvests, many weeks after the new yam festival, I used to see men visiting my father for an opportunity to see and admire his barn of yams. The barn used to be a big spectacle in my village, including some gossips that went with it. In the barn, you saw large yams tied in columns starting from the ground and reaching heights of about twelve feet. Children used to run around the barn happy.

But now, this type of sight has disappeared. Wretchedness took its place. Why? You may ask. Did the Creator's bounties disappear? No farmlands, crops, seeds to sew? No sunshine, adequate rainfall, and even the air? Of course, these gifts were and will always be there. So, what is the problem? What changed?

The answers to these questions will always be that nothing changed. The Creator's bounties remain as there were. Peoples' attitude towards the Creator's instructions is the common problem! All living things must comply with the Creator's law – 'go and get' or suffer the consequences as evidenced by, in this case, poverty. Praying any number of times per day will not blot out the stains of disobedience to God's commandment, 'go get' the food I have already provided for you.

No! People will instead pray. Now we have prayer houses—churches of different denominations with one in almost every corner, and depending on the region, you may see mosques, temples, and synagogues. Everybody belongs to one of the many varieties. On any given worship day, people can spend hours praying and worshipping. Praying is much easier to do than physical work. Ask God for anything; you are 'sure' to get it, they believe! Really? The most confusing part of this ideology is that these people have been 'asking' for years and centuries but have not received anything or seen betterment in their lives. It is an addiction and unfortunate!

It is interesting to find out what happened to the belief that all one needed to do was ask God for anything, and it will be granted? These are simply words without substance. And the unfortunate part of this creed is the donation requirement – the obligation to make donations regardless of economic conditions.

Visit any poor communities or countries of the world and observe the fervent prayerful worshippers. The situation is always the same. People in these places abandoned the Creator's prescription to live on this earth to pursue human instructions.

Consider a common saying in many places of worship, "You don't give, and that's why you are poor." Just think about this statement for a second! It is implied here that when one gives or sacrifices to the Creator, will He reward the individual in multiple folds? Under this mindset, one can be persuaded to give up his last dime.

Does this not show total abandonment of the Creator's gifts of the brains, senses, conscience, etc., He gave to humans to help them steer through the rough waters of life – the ever-present deceit? What contributions or how many hours of prayer did humans make to persuade God to provide the sunlight, the food we eat, water, the air, fertile soil to plant on, beautiful things of this earth that caress our senses, etc.? None! After mortgaging one's free will to the religious leaders and consequently being taken for a ride, the Creator is blamed for standing by and doing nothing! This foolish idea leads to hopelessness and despair.

But the Creator commanded all living things "to go get their food." Does this instruction in any way mean to sit idle and wait for handouts or pray? Their systems are so designed that falling into poverty (as we know it) would not be possible in developed economies. All non-disabled adults ready to work can usually find work to do. For the educated, employment is almost assured.

At this juncture, let's examine the Creator's provisions more in-depth. It will be instructive to define what is possible or impossible for humans using the resources from the Creator. Any issues about sustenance, protection against predators, living things existing here on earth, the ability and means of taking care of our earth are all provided by the Creator.

But some events are outside this group – things that happen beyond one's control using the Creator's provisions. I'll cite a few examples to explain: one is on a flight, the plane catches fire and explodes. What if being in a crowded place where someone detonates a bomb or explosives? There are events like hurricanes, tsunamis, earthquakes, etc. In these situations, one must do what one can, using the brain and senses given by the Creator. Then resign his fate to the Creator, knowing that there are evil people on our earth.

We may have witnessed miracles in such unusual situations: in an earthquake, some people are pulled out from the rubles many days after the event; people survive deadly car crashes; a dolphin aids a drowning person, or a person survives a shark attacked with little harm, etc. Remember the saying, "I am your Creator who provided all you need, and I'm always with you."

Going back to the discussion on survival using the brains and senses, we will examine the case of a group generally described or seen as the 'unfortunate of the earth.' They live in remote villages and very rural areas. They have no one to depend on but themselves.

You may wonder why this group is called 'the unfortunate.' It is because they live in the villages with their wives, children and grandchildren, parents, and grandparents; their main occupation is the looked-down-upon farming on which their livelihood depended entirely. Remember, their agriculture is not the modern 'hi-tech' type but the hoe- machete-manpower sort.

During planting seasons, they work from dawn to dusk. To them, farming is a family affair; it is a way of life. They produce different food items, including various types of vegetables, fruits, seeds, tubas, meats, eggs, etc. They may sell some of the fruits of their gardens like the yams, vegetables, palm oil, and palm kernels to raise money for other necessities they could not produce like salt, clothing, etc., and send their children to school.

Imagine the type of land they planted on – virgin soils untouched by modern chemicals and insecticides. It is assumed these people do not know better, but they practice crop rotation. "What is the quality of their products?" you may ask. "Just as the Creator intended!" Their food is untouched by civilization. The seeds, fruits, vegetables, livestock, and soil provided food, nutrients, and medicines as the Creator intended. Sickness or what we see as normal ailments are generally absent. Can 'we' really believe this?

During planting seasons, these 'unfortunate' peoples work very hard, sweating out toxins in their body, exercising, or keeping the body in constant motion as the Creator designed it. What of the free vitamin D? Their bodies are always soaked in nature's free sunshine.

During and after dinner, they have all the time to enjoy the company of one another. Why? Because they have no large bank accounts or stocks to worry about, no fears of any person coming to their houses to cart away their jewelry, expensive property. Remember, they have no mansions in every capital city of the world, boats, private jets, or even fleets of cars to think about after dinner. They do not need to fly overseas or engage in long-distance journeys. There is no need for those?

Then what do they do during and after dinner? Enjoy the presence/company of the members of their families – father, wife, children, grandchildren, and grandparents! What can be higher or more enjoyable than this!! Who should call who "unfortunate"?

Now, the city man or the educated person has begun to learn some lessons from the country person: many are turning to organic foods. The country person does not need to worry about such expensive produce. The country person does not need to pay for fitness membership as the body is in constant motion. He is very healthy, lean, with no overweight problems, no arthritis or cavities or tooth problems. He does not need to sit throughout an eight-hour job in an air-conditioned room with no opportunity to see or be in the sunlight.

Many people are ridden with different types of diseases – consequences of disobeying the Creator's laws: 'Go, get the food I have already provided 'Each word' in that instruction needs to be scrutinized and understood. Each word has a profound meaning. "Go, get the food . . ." does not mean: Sit on a chair for hours in an air-conditioned room and when hungry, order a take-out dinner. Remember, 'Go' involves movement – exercising the body; the food He is talking about is the kind He designed for our body, not genetically altered or produced on lands contaminated with chemicals.

Remember also; the Creator is all-wise and all-knowing. All processed food, all packaged food, all packaged food, all GMO foods, etc., are regarded by our bodies as foreign materials. They become poisonous when they enter our bodies. The consequence will be inflammation, diseases, sickness, and death—unhappiness.

There should be no sadness in a place the Creator designed to be enjoyable. But because humans refused to follow the Creator's instructions, unhappiness entered the earth! Indeed, the education we received about the Creator was wrong.

Assimilation into an alien culture is not easy because it involves mastering the workings of the new culture: the language, changing one's natural accent, diet, sleep pattern, etc. Having a different skin color does not make assimilation easy. It allows the majority to point out the 'aliens' even if they have mastered the culture. Additionally, the alien has to deal with constant challenges of natural perspectives or worldviews. These will always identify or expose the alien and cause ongoing conflicts and discomforts.

Previously, I briefly discussed the justice system in the West. In my new home in America, I have watched the Justice System at work in the media and the courts. It was rather alarming; I rarely witnessed justice in the justice system. What I experienced was always the party with more experienced, more knowledgeable lawyers and money to pay them, always won. Remember, this is the culture, and nobody in the culture quarrels with it. It is accepted as the way it should be for them.

Also, I mentioned that I came from a culture where it is the parents› duty of a person who committed a crime to call out their person to face or pay for his crime.

The parents were not supposed to protect or encourage crimes. This requirement was how it had always been until recently, when lawyers and the influence of money and religion forced themselves into and corrupted the legal system there.

Are two cultures being compared here to determine which is better? It is like comparing apples and oranges. None is better! Both are good – but suitable for its people. Rendering justice in situations involving aliens or people in the process of assimilation to such a culture is challenging. How will such people find a jury of their peers, for example? As explained earlier, in many European countries, including America, an accused person may not be questioned in person and rarely represents himself but must have legal counsel or a lawyer. One is even allowed to take the '5th'. All these provisions were to avoid self-incrimination! To adopt this method lessens the impact of the conscience and sense of responsibility. Therefore, it's easier for the person with the help of his lawyer to wiggle out of a crime.

How will aliens to this culture, unfamiliar with the justice system here, feel? The problem here is moving or being forced to move permanently into an alien culture. The law is not the problem because the country's citizens established it, and they are happy. The inconveniences that emanate from moving or being forced to move permanently into another country show that it was not the Creator's intention. All living things have their Creator-assigned places where they understand the culture and thrive in it. But if one decides to move permanently to another culture, he must try to understand the culture first. No culture can change simply to suit an alien!

In the Middle East and some other parts of the world, it is considered an outstanding achievement for people to marry their cousins. In Nigeria, that would be an abomination. But is there anything in nature that suggests that was not the intention of the Creator? Nothing! What should be observed here are differences in cultures. Cultures are not supposed to be the same. And there are deterrents already in place to prevent any event against the Creator's will from ever happening.

We know God designed humans and endowed them with brains and conscience. He bestowed upon them the ability to choose, make determinations, pass judgments, and impose consequences as they impartially see fit. We also know that there are different ethnic groups on this earth and each with its peculiar cultures. Again, which culture is the best or most superior? None! All are good for their peoples. So, God of infinite variety allowed them. Nothing is one except the Creator.

It is true; all cultures may have some flaws or harmful practices that may need changes or modifications. The change must not be imposed from outside the community; it must come from within, from the people themselves. There was once an uproar over the circumcision of females in some parts of Africa. Some people, particularly feminists, argued it was inhuman for parents to subject their daughters to such a 'barbaric' act that would rob the child of sexual pleasures and satisfaction at adulthood.

The parents, particularly the mothers, disagreed with this view because they felt that it was a practice that started with the time and saw nothing wrong. The mothers accepted the practice because it reduced the erratic sexual desires in some females without circumcision. They argued that the consequence of such behaviors was an unwanted pregnancy, abortion, diseases, and family shame.

Many parents have abandoned this practice because they understand more about the harmful effects of female circumcision and are slowly acquiescing to the dictates of modern thinking. You can imagine the amount of pain and anguish the practice generated. All glory to the Creator.

In conclusion, culture represents the people that developed it and allowed by the Creator.

Therefore, supplanting a culture will always spell disaster. Some cultures may need some tweaking; that change must come from the people, not imposed from outside.

Chapter 13

PRAYERS

We can only imagine the type of gratitude shown on the face of the 'first' farmer when he saw that his crops had started producing food for his family and fruit trees had also begun producing fruits, etc. He must have been thankful to the Provider. But now, with what, and in what language did he do that? We can only infer from what makes sense. How does a person express gratitude? In different ways! The most common ones are verbal – like thanking, praising, commendation, even returning the favor or material objects – offering money, providing some form of services, food, clothes, animals, etc.

How did humans express their gratitude when they saw all the good things provided for them by the Creator? In addition to the food, the sun was available, keeping them and their animals warm and helping them germinate and grow their crops and ripen their fruits; their food was everywhere, and the soil was very fertile. Water for drinking, bathing, cooking, and for their crops, plants, and animals was in plentiful supply. The life-saving air was everywhere. Most amazingly, all were provided free and in abundance. Their gratitude was undoubtedly great.

Scientists had said that the Creator had provided these human needs millions of years before He put humans here on earth. This statement will make sense only if the Scientists believed that the time was as the Designer desired, but not to support the "Theory of Evolution" — that humans evolved from Apes." If that theory was correct, then there was a need for such a length of time for Apes to transform into humans. Has anybody seen an ape turning into a human being? The core foundation of that theory is flawed because there was no instance, from the beginning of the recorded history to the present, that anybody saw apes changing into humans. Or was the transformation in effect for a time and then stopped after enough humans (Adams and Eves) came into existence? Nothing in our physical world shows that ever happened.

The Creator, the All-Knowing, does not need that length of time. He wills, and it happens. Otherwise, how long would it have taken the Creator to 'make' the stars, planets, in short, the universe?

Back to prayers: They must have used some verbal expressions to thank the Creator that provided all the good things for them, as is usual with humans. But in what language was the verbal expression of gratitude made? It is most reasonable to assume it was in the language of the person expressing gratitude.

The Maker, a spirit, does not need to hear, listen, or prefer a language. The Omniscience, who knows our hearts and minds, need not speak any language. Only humans need to acquire or learn languages and communicate in them. Since time is for human convenience, the Creator knew what His creatures would do before they came into the 'earthly' existence. Does the car designer not know all about the car: the speed, the power, etc., the car will have after its design and construction? The Fuss about talking, praying, or communicating with God in a language was a human conception and manipulation.

The same reasoning goes with the practice of offering material things to God as a show of gratitude for His provisions to humans. It makes sense that a farmer that produced potatoes or yams or the person that raised cows, goats, or chickens, will offer the best from what he had. It is logical to believe that a farmer that cultivated tomatoes used tomatoes, not chickens or cows, he did not raise. In other words, he used what God blessed him with for the offering. Also, we know that the designer is not human and does not need material things. Each person, each community, used what is available in that community. Japanese, British, Ghanaian, etc., will use what is available in their communities. A British, for example, will not use yams since yams didn't grow in Britain.

It would not make sense to brand any sacrificial material superior to others or the only type acceptable to God. God is a Spirit and does not need us to offer back to Him his earthly creations. Therefore, the idea of God favoring or preferring one type of sacrifice or worship above others should not even arise. Humans indeed constructed the picture to advance their superiority over others.

People who made burnt offerings then believed that if the smoke from the burning rises, it meant God accepted the offering. Remember, smoke a lighter gas must always rise. And God, a Spirit, does not need or breathe in any smoke.

Displaying or expecting gratitude is human and animal behavior and in this world. Since God is a spirit, and as such, does not eat any form of food or breathe in smoke from the burnt offering, why do we believe He expects or demands sacrifices from humans? The Creator does not wish or require or have any need for any of these different forms of showing gratitude.

These forms of gratitude benefit the person showing appreciation, not the Creator. It is, however, very reasonable to believe that humans need to show gratitude to God for everything He provided for them, including their ability to enjoy those things. Simply because it is a part of human behavior, it is suitable for them not to forget that their whole being depends on the Creator.

Since using material things to show gratitude did not make sense to everybody, some people thought they could do it through prayers and constant 'devotion' to God. Following this line of thinking, many people spend their whole life in prayers and worship. Others exploit the idea to enrich themselves.

Perhaps you have experienced a session of worshipping in which 'making joyful noise is practiced.' You will be carried off your feet by vibrations from the singing and drums. You will see people profusely sweating, and many had fallen on the floor, foaming in their mouths; those speaking and praying in tongues will give the impression they were in the do-or-die competition for the first position in loudness.

These practices can, at best, provide some satisfaction and stress relief to the practitioners while these are going on. Often people who have this emotional/ spiritual release can continue their daily responsibilities sanely, with no need for counseling or mental health treatments. This intrinsic value is perhaps the most beneficial aspect of these practices and prayers!

I have read about the Essenes of Egypt, visited some monasteries, worshipped in some Mosques, and prayed with different Christian denominations. I can claim I have experienced what these institutions are all about. Also, I have watched my father and mother pray. I observed one common denominator: All prayed directly or indirectly to a higher (highest) power – the Creator God. The way one 'approached' God depends on where a person sees oneself to God. Some people believe they were inadequate, unworthy to approach God directly. This group 'went' to the Creator through an intercessor.

My father and mother (in their native religious beliefs) called on the spirits of their dead parents, the Spirit, Ala (land), and other forces in nature to carry their messages to God. They felt they could not approach the Maker directly by themselves due to a feeling of inadequacy. Some Muslims go through Prophet Mohammed (despite being told not to), some Christians go through their Prophet, Jesus. Chinese go through their Prophet Buddha, etc.

A small group believes that since God is everywhere, 'in' everything,' and closest to them than any other thing, they don't see the need for an intercessor. I don't see anything objectionable with any of these methods because God alone knows the intentions of humans and the degree of their spiritual growth. Therefore, it does not make sense when some humans assert religious superiority to others to claim that their prayers and practices are superior. One standing before the Creator depends only on how much one understands who the Creator is and obeys His laws.

Before further exploration of prayers, let me define them.

What is Prayer? Does God need prayer, worship, or devotion from human beings? Prayer is a humble way to show gratitude or to ask for a favor. Asked differently: 'What is the intention for praying?' Peoples use prayers to express appreciation or ask for needs. So, when we pray to God, we humbly ask God for guidance, provide whatever we lack, or thank Him for what He had already done. Humans implore God for help with something they lack, desire, or which is unavailable to them.

The help they are asking for could range from immaterial guidance to the provision of material things like money, food, etc. Many people believe that through prayers, God will grant them whatever they are requesting.

Is this not like saying that humans can affect God's intentions and decisions, i.e., prayers can influence God somehow? Nothing in His creation, our physical world, shows this is true. How can humans affect or influence the Creator's intentions or decisions?

Please consider the following: The universe, including the earth, was completed millions of years before humans were created, as legend assured us. It would be reasonable to believe that since God is the Almighty and the Ever Knowing, the length of time was as He desired, not that He could not do it in a shorter time or needed any time at all. I don't believe that God needed such a long time to create and put human beings on earth. God cannot make a mistake; He can't forget. Time-limited Humans, created by God, are, therefore, in no position to question or influence God's decisions. How can the "little inconsequential" human beings control God's will?

In the previous chapters, I described the mightiness of Almighty God using numbers and distances: The number of stars was in a million billion, and the distance between our Sun and the nearest star to it is about four light-years. And there are billions of galaxies, with our own Milky Way Galaxy being one of the smallest. Remember, these numbers are approximations because scientists are still discovering more galaxies.

When you think abo these d stances and numbers and consider the complexities of human bodies and other an s, you can see that the human mind cannot even begin to comprehend the mightiness of God. Our brains and senses are too limited to handle such knowledge. Can you then imagine a human being, with no understanding of divine plans, the most disobedient of God's creations, talking to God, telling Him what to do!

Many times, I'm inclined to think that it is arrogance and pure ignorance and disrespectful for humans to believe they can talk or make suggestions to God not to talk of influencing or altering God's will to suit their own. When one ignorantly thinks he is 'talking' to God, it's like imagining things or daydreaming. Some say you can talk to God in spirit. But humans are earthly and, therefore, incapable of speaking in spirit. Only spirits can communicate in spirit. Can you imagine a human being standing before the Majesty of God? Then, one can also physically touch the Sun!! Think for a second about the absurdity of such a notion!

When a person talks to another person, they generally exchange ideas, ask for favors, etc. How can humans exchange ideas with God, the Creator of the universes, and all therein? Humans try to change, alter, or even 'improve' upon something that has a fault or isn't complete. God's creation is comprehensive and perfect. How can humans make suggestions to God or remind Him of anything? Can He forget? God designed humans and therefore understands the needs and workings of humans more than they. How can any person ask God for a favor or ask Him to satisfy a need? Before that person was created, the Creator had known all their needs.

Some people have some bizarre notion that they can get God's attention if a prayer partner, a group of prayer warriors, or many people are joined to help them out. This thinking is another nonsense. Does this imply that God is deaf, or you need more people to persuade God since more people will exert more pressure than one person? We all, including those who peddle these irresponsible ideas, know that God does not need persuading.

Why would God need humans to make a plea to Him on any person's behalf if He already knows the innermost thoughts, desires, and intentions of those people and has already taken care of the needs? Can anybody point to anything in His creation, in our environment that shows He desires this? Nothing! This type of thinking shows humans have confused ideas about God's infinite powers and knowledge. And that humans got it wrong, right from the beginning or instead were misled by their religious teachers.

Don't get me wrong or misquote me: prayers have their use, depending on the content of the prayer. First, prayers are for our good, and we should do that more often. The problem with humans, the prayerful, is a lack of understanding of what prayer is and its purpose.

Consider the feeling after what is considered a fervent 'prayer' to God – in which one resigns oneself to His Will, acknowledging one's unworthiness for all He has done for humanity and that without Him, nothing exists. That would make one feel good, happy, and satisfied. Any other intentions or form of prayer would not make any sense.

Any person who prays (in any form) with the primary purpose of telling, dictating, or reminding God what he wants God to do for him should realize that God does not need to be coaxed to do anything. Besides, all that His creatures needed were known to Him and already provided.

Humans should instead 'pray' (talk) to themselves whenever they need to pray arises! Why and how? Humans are incapable of communicating with the Almighty, the Creator, in spirit as they claim. Humans are earthly and, therefore, not spirits. God is a spirit.

Also, it would serve humans better to 'pray' to themselves, i.e., remind themselves who created them and have already provided everything they need with nothing missing and that without Him, they don't exist. At the same time, they should remind themselves of their duties here on earth or why they were created in the first place.

When you wake up in the morning, would it not be better to remind yourself (pray to yourself) of your duties here on earth – the reason God created you and instruct yourself to strive to fulfill them. And before you go to bed at night to question yourself whether you achieved all that was required of you that day and affirming to try harder the following day. These are all the prayers we need.

Think of the notion that the Creator is in everything! This notion has led many to believe that God is inside any house of prayer. They call it a holy place. Even some go as far as believing that God exists or locked up in pieces of bread, they call "holy communion." The Mighty Being that created the universe, including all the unseen things, can 'measure' the incredible distances between stars, can't be bounded by space.

Is it not madness to believe, even for a second, that the Mighty Creator can be contained in a piece of bread or house? He created everything and was not bounded by time and space. Therefore, He is everywhere and cannot be contained in any materials, least of all things made by man. Yet, He keeps His 'watchful eyes' on all things. Or else, many things will go wrong. The notion that humans can invoke or command the Creator to reside in a house or a piece of bread, earthly things, does not make sense.

We can verify the above statements by going back to and examining nature – other living things! Birds, plants, trees, fishes, worms, elephants, lions, dogs, chickens, etc. They do as the Creator commanded – they grow, multiply, produce the food or shelter or medicine for all living things, live and let live, as they were created to do.

They don't make their laws to spend hours or days praying, fasting, going on pilgrimage, involved in holy wars, carrying on, giving the impression they are doing anything good. If the prayers mean anything good or anything at all, why all the killings, unhappiness, cheating, selfishness, etc., we witness daily? Or does it mean we don't pray enough?

Is praying aligning oneself with the will of the Creator? Yes. Then, what is the will of the Creator? Since this subject has been discussed many times in this book, I will address it here differently in the following format:

Just imagine for a moment how enjoyable and happy our earth would be if: humans were to wake up in the mornings reminding (praying to) themselves of who created them and what their duties here on earth are. They should strive to accomplish their tasks and assure themselves that all their needs are already known to and taken care of by Him. The world would be a paradise.

Also, He has shown us the way or provided us the resources to take care of our needs. And humans should realize that they are nothing without Him! Above all, they should understand that the Creator gave them the brains, intellect, the senses, the conscience, etc. to aid them to carry out His will – all living things will live here on earth forever, and humans would take care of the planet. Finally, our world would return to the Paradise it was created to be. Suffering and unhappiness would disappear. Our existence and all we need to sustain us here on earth are from God. And God knew all we would need before ever we were created. Does the baby on the laps of its mother, being breastfed, ask the mother, "Mother is the breast milk you are feeding me clean, warm enough, palatable; can I trust your judgment that you know what you are doing?"

No! The baby instinctively knows that whatever the Mom does for it must be right. No questions asked. The mother theoretically knows the needs of the baby, including the time to feed it. Humans should learn from babies and have the same attitude towards the Creator. Many holy books like the Bible emphasizes that humans must think like babies in their relationship with God if they ever hope to go to 'Heaven.'

Secondly, prayers provide guidance, particularly to children and their likes who don't quite understand who God is, His powers, knowledge, greatness, and magnificence. By explaining to a child what God provided to humans and human's total dependence on Him, the child can be humbled and ready to obey God's laws – to care for one other and take care of all His creations.

But think of what children are generally 'fed' about the Creator – lies. For example, because God loves you, He will forgive any sin, no matter how many times it was committed, or pray five times a day, you are good to go. Moreover, they are taught more about spending their time doing rituals than their first duty to take care of themselves, each other, and the earth.

I emphasize that humans did not get it right in the first place or were not taught right about who God really is, what should be humans' relationship with the Creator, and the purpose for which He put humans here on earth. Or were humans somewhat misled by their religious instructors?

There is no other way to explain that humans don't understand who God is and, therefore, do as they please to indulge their appetites. The unfortunate consequences of such careless and ignorant behaviors are already built into God's system. Refusal to accept or understand what God requires of all of us in the name of free will is no excuse. He left us with the brains, senses, consciences, etc., to make our choices. For choosing to do our intentions, and even worse, to willingly subject ourselves to the will of other humans continue existing in their self-inflicted agony until they voluntarily return to God's order.

Imagine the amount of time spent praying, reminding, coaxing, or even trying to force God to do according to our wills—on matters He has already known and decided? Haven't you been in a situation where you were expecting the worst to happen; you were sweating, nauseous, and had a sleepless night? But low and behold, when the 'doomsday' arrived and the hour came, nothing happened.

Do you remember the surprise you felt, almost ashamed of yourself – your doubts, uncertainty? Yes! Because the Creator had known about that situation and taken care of it even before He created you. Whatever is out of control of humans (they are indeed many), the Creator takes care. Think of the peace, the love that would reign on this earth if we recognize that God has already taken care of future situations that give us anxiety. All we need to do is to know Him as our 'Parent', trust him like a baby on the laps of its mother being breastfed, and use our brains, senses, and resources he had already provided to us to make choices.

Trust in the Creator is the most crucial aspect of the relationship we should have with God. Many times, things don't turn out the way we would prefer. But our ideas and the Creator's are not the same. All we need to do is trust that the Creator knows everything; whatever He does is for the good of humans, and He provides appropriate and adequate resources that would enable us to make the right choices. Then, the stresses generated by the fear of things that are out of our control will be eliminated.

It is merely human behavior to show gratitude for any form of kindness done to them. It is also human to expect a display of appreciation from a recipient of some good deed. Can humans do anything for which God should be grateful? No! It is unreasonable to believe that God designed humans to worship and praise him. How can humans, earthly beings, do that to God, a spirit? So, God does not need all that.

What do humans have that God needs? First, whatever humans have, God provided. In human terms, thinking that God needs all that from us is like giving someone a basket of apples from your apple tree and demanding that the receiver thank you by giving you back some of the apples. That would not make sense.

When a young person or any person does something good, they expect something in return. If a young person cleans his room, that would please the parents and attract some reward. If a person sends a child to school, he expects the child to grow up into a valuable member of society; be able to provide for or help himself and others and have better self-esteem for the achievement. The parent or sponsor would feel good about having done the right thing.

Is it possible that God created humans and provided all they needed to live and enjoy themselves here on earth, and free of charge because He wanted to feel good about Himself for doing the right thing? Not at all! Feeling good or feeling bad is earthly. Then the Creator must have His reasons for creating humans if it was not for them to pray to and worship Him. What might that be?

It has been identified previously in this book that God created humans to take care of the earth. That was the conclusion reached after examining what God created and the place of humans among all creations. There is no other way to answer that question since no one can read God's mind or ask him directly for the answer.

Here are a few takeaways from the discussion on prayers. Praying is for the good of the person making the prayer. Also, God does not need any form of worship, talking to, persuasion, coaxing, etc. Humans should instead pray, coax, persuade, or talk to themselves in times of need, danger, or when they are simply down and remind themselves that the Creator knows everything and then use the resources, He had already provided to act on solving their problems.

What should be the content of this prayer? First, acknowledge the ever-presence of God; that He created us and knows all we need even before we were created; trust that He had already taken care of and made provisions for those things that worry us. How long should praying take? One minute, one hour, or weeks?

That will depend on the individual needs, understanding that God does not need any form of praying from us. All He wants from humans and other animals is clear: read and understand His intentions as 'written' on His creations in our environment. Then we need to persuade ourselves to calm down (meditate, quieten our minds) and be patient and trust that the Creator knows and has taken care of those problems that are out of our control.

There are many desperate situations wherein we find ourselves attempting to persuade or even command the Creator to act according to our wishes. That is childish and demonstrates doubts in the Creator. What about trusting the Creator as the contented baby, described earlier? Many of us, unfortunately, fail woefully in trusting our Creator, the Almighty God.

Hence humans are forced to resort to unnecessary actions that rather question the existence, power, and knowledge of the Almighty God, resulting in incurring retributions.

It is not uncommon to hear people claiming that they saw God or had a conversation, or received instructions from the Maker. We know human eyes cannot see a spirit. Granted, the Creator has the infinite power and knowledge to do as He wishes to inspire humans. Otherwise, how can humans, earthly beings, see God? How can a human being (in human physical form or even in 'spirit') stand in the presence of the Maker, before the Maker's Majesty and Glory? How can this be possible? The creation is complete. Even the things that will happen in the 'earthly time' are already known to the Creator. And he has taken care of it.

What of prophets and 'holy' people, those that claim special favor from God? That does not change the reality. Those humans with unique talents – genuine visionaries, geniuses, painters, musicians, engineers, doctors, etc., were specially created to bring some information or knowledge to living things.

This did not happen because the Maker, at a point in time, remembered that something was missing in His creation and decided to send people to complete it. The Creator can't forget. All we call 'changes' or 'amendments' in God's creations are already in the Creator's blueprint and brought to human consciousness in the time accorded with His plan. Remember, the Creator can't make a mistake. His work is perfect and complete.

Humans can only gain the knowledge mentioned above of God, the Almighty, the All-knowing, through studying His awe-inspiring perfect creations around them. He does as He wishes. Occasionally, some people are of the impression that the Creator, from time to time, must tweak His creation to suit the current times. Humans forget that time is earthly. Anything that has happened or is happening or will occur in the temporal time had already happened with the Creator.

Humans are inclined to ask where God was when these events were impacting them in times of great calamities like death, earthquakes, fires, etc. Some go as far as asking, "Why do 'good' people suffer?" Humans should be most concerned with the events that humans, in some ways, brought on themselves. Let's examine some examples:

We know how fossil fuels came to be. Is it possible that the Creator of all the good things of this earth also created fossil fuels to enable humans to use them to pollute and poison the atmosphere? Of course, that was not His intention. We can't forget that the Creator gave humans the brains, the senses, and the conscience to take care of the earth. The problem here is the failure of humans to use these resources.

Everything created is for a purpose, for the good of living things. Even the venomous snake has costly leather (because of the unique designs and skin toughness). Also, it is food, and even the venom has medicinal use.

The brains given to humans were to be used to unlock and discover the good in every creation. This must be the way the Creator intended – that human, bounded by time, must spend some time and energy to discover the good in and purpose for every creation. Or else He would have unlocked all the secrets and mysteries Himself. God so willed it. We don't know what He knows. We are in no position to question Him.

To understand and measure the efficiency of a car, one must understand the design and workings of the vehicle. All we assume to know is that humans are created to be, among others, the earth's caretakers, use the earth's resources for the benefit of all creations. That's how far we can go.

No human can understand the thinking of the Almighty Creator and the design and purpose of the 'human-machine.' And no human can evaluate or judge His work. If humans can't understand the purpose and meaning for any creation or any event by examining other expressions of the Creator, they are left with no other choice but to take a 'leap of faith' that all He created was for the good of all living things. If we eliminate human activities that carry consequences when the Creator's laws are violated, we should accept any event as so designed by the Creator.

If a child is born bisexual, gay, tall, short, etc., would these be attributed to human activity? The answer is a capital 'No.' But if a child is born with traces of cocaine in its veins, that certainly would be attributed to the parents' activities.

Chapter 14

MISGUIDED SAYINGS

We often hear the following expressions: My God is good and merciful, but yours is unkind and wicked; My God is kind, He will forgive me if I repent and ask for mercy; say the name of Jesus and all of your evil acts or sins will be forgiven; I'm a favored son/ daughter of God because the blood of Jesus washes me; I desire to acquire the 'key' that can unlock the gates of heaven; Pray and ask God for whatever you want, and He will grant it.

Sometimes humans think they can entreat/force God to do what they want. Let's carefully examine these few examples of misguided beliefs and understand why there is so much discontentment on our earth originally designed for perfect happiness.

Does the sun not shine on us all? Are we all not given air to breathe? The words kind or cruel, good, or wicked, evil or merciful are human expressions used to describe human relationships and behaviors. How can the Creator of all things be characterized as being kind or cruel? These two dispositions define the ability to favor or unjustly deny or withhold a favor or refuse to deliver as required. They are not characteristics of the Creator.

A kind person can forgive easily; he can look the other way when certain offenses are committed. It is easy to overlook a beloved child's shortcomings. These behaviors imply accountability and discrimination. Whatever we do here on earth – good or bad – involves accountability to some authority, like the conscience. This commandment gives humans the opportunity to amend their misdeeds willingly.

Even when a misdeed is done in secret, the ever-present conscience witnesses the act.

It can amazingly pass judgment and deliver consequences immediately. These willful misguided sayings of people do not pertain to the Almighty God, the Ever Knowing, who stands alone by Himself, requiring no help and does as He pleases.

When it rains, or the Sun shines, or plants produce fruits, there is no element of favoritism. Who or what authority can judge God? To decide, one would have to compare, using the details for comparison. There is nothing on this earth with which to compare and then judge the actions of God.

We should simply be like the baby, described above, on the lap of the mother being breastfed. The baby is completely satisfied and has no reason ever to doubt the sincerity of the mother. Humans like to find the easy way out, avoid responsibility, and appear perfect and blameless before their peers and neighbors. Hence the use of such statements as My God is good; I'm a child of God; He will forgive all my sins; the blood of Jesus washes out my transgressions, etc.

Some people even believe they can take their transgressions to another person who has the power to forgive sins or the ability to persuade God to forgive the evils. Is there anything occurring in nature that indicates that a human being has the potential to forgive sins? There is no such thing.

"You don't have to obey the laws of God but need only to say the name of Jesus, and he will forgive all your sins. If this is true, why then do we have so much unhappiness here on earth? No human nothing can alter, change the Creator's will or persuade or dictate to God, Almighty. God is God, the Arbiter, the Beginning, and the End.

Those who parade themselves as having powers to forgive sins or the ability to dictate to God or influence His decisions are, at best, misguided. Remember the duties of all living things and that of the humans in particular – sustain yourself with what God has provided; protect yourself and what belongs to you from predators; extend your kind through procreation and protect all living things.

The true religion for us earthlings is to be conscious of and diligent in our responsibilities to each other and all creations. The Creator provided to all living things the means to achieve all things. Worship, therefore, is appreciating all that God has given to His creatures and reminding ourselves of the Creator's omnipotence, omniscience, and our total dependence on Him.

Another bizarre belief is that a human being can issue the 'key' to 'heaven.' Some adherents believe this literally. With this 'key,' they assume, the opportunity to enter heaven is assured, the state of the 'soul' notwithstanding. To acquire this 'key,' a person can be induced to murder, destroy, or commit all sorts of evils.

Some religious leaders have used this deceit to entice young people to do their bidding – commit all sorts of crimes. The most disgusting part of this scheme is that many responsible and religious people stand by while this evil is being carried out.

The place called 'heaven' cannot be earthly because it is believed to be the abode of God and His servants, the angels. How can a physical material, 'a key,' be able to unlock it, even figuratively? You begin to wonder how any sane person can believe such a fallacy, particularly when you consider the gifts of the brain and the senses.

Another saying is: "I love God." I once had a conversation with a friend because he asserted that he loved God very much. I asked him to explain what loving God is. The look on his face indicated he was surprised at my question. But I persisted.

Then he began to explain that God created everything and provided all we need here on earth because He loved His creations. But all that talk did not show me how he loved God or explain how to love God. Finally, he said, "I am a servant of God. I go to my place of worship regularly without missing a day. In any day of worship, we spend several hours praying and worshipping." "Do these not show I love God?"

That prompted my next question, "How do you show you love your wife?" He proceeded to go through a litany of things he did because they pleased the wife, like caring for his wife's needs as he did for himself and doing everything that made her feel good and happy. "Those were loving and considerate acts," I whispered to him. And when that lucky woman was not home – out on a visit or traveling, the gentleman kept his wife in his mind and restrained himself from messing around. And when she returned, he made sure she noticed how much he missed her presence.

I thought that was beautiful, and we agreed those were ways to show love. Then came my last question, "How would you show love to the owner of the house you were renting? Suppose you did not know and had never met the person before?" He hesitated and then said the question was a tough one. Why? "How can you love someone you don't know, never met? "I would take good care of the beautiful house," he exclaimed. He was not sure whether taking care of the house was because he loved the landlord or the building. That took us right back to my original question about loving God. Love is earthly and an attribute of all animals. To love, you must identify the object of your passion or the extensions of the person. It is born out of a human need for endearment. The only way for humans to demonstrate love to the Creator is through caring for and appreciating all the things He created. And humans must always understand that it is the Creator that created all things. The understanding and appreciation will encourage our obedience to all God's commandments, including natural order and universal laws, notwithstanding endless opportunities for our scientific inquiry and discoveries.

No one has ever seen God before, only His wondrous works here in His universe and our imaginations of Him. People have written books about God. All are the imaginations of the people. Even the 'inspired' or 'revealed' scriptures require our belief as we did not witness God's 'revelations' to the prophets or writers. God is God. Humans should not doubt what He can do.

Where can anybody begin to define God? Human knowledge is limited to how much we can glean from His creations. Our relationship with the Creator is in no way the same as the relationship between my friend and his landlord. God created all living things and made them what they are, and without Him, everything will cease to exist.

Humans and, to some extent, other animals have the need to be loved and to express gratitude and love. Since we can't see God but only the expressions of himself as in His creations – we can only show our appreciation for Him through loving the things He created. But how? Humans can do this by caring for and protecting all He created. Doing those things is the only way to show love for the Creator. We must do these things because our fate and those of all creations are tied together in an everlasting bond. In fact, to do these are commandments with severe divine consequences for failure. Loving all creations is for the good of all living things and not for the Creator.

"I will pray for you," or "my thoughts and prayers go with you!". What is the value of this statement? Does the idea make any sense, or is it merely an expression for lack of something better to say? It is better to answer this question with a real-life story.

While on her way to visit a friend, a woman met a young man who appeared familiar. She was surprised because she was not from and never lived in that city. On the face of the young man, she saw misery and poverty boldly written. Her belief that she somehow knew the man and the miserable state she found him persuaded her to approach and strike up a conversation.

Low and behold, the young man was a relation to her high school boyfriend, and now a man of wealth, who could boast of at least a mansion in their capital city. You can see why the woman was surprised at finding out who the young man was. Without prompting, the young man started narrating his encounters with hard luck.

'You can see I'm young, strong, and love to work, any job that can help me meet my family needs,' He told her. 'Yet I can't afford the school fees for my two children,' he concluded. "From all you told me," she said, how come you can't afford your children's school fees, she queried. Besides taking care of my family, I also take care of my two deceased brothers' families and that of my sister, who her husband abandoned.

"Must you take care of these people; you are only responsible for your own family," the woman lectured him. He replied, "I know, but I'm not so wired; my brothers would do the same for me if I were in their state." The woman asked, "Is your rich relative not aware of this?". "He is," the man replied, but he offers nothing and only leaves his 'thoughts and prayers' with me. "You can see, nothing has changed," he lamented.

"Thou shalt not kill"! Whose law is this – the Creator's or humans'? What does it mean; you should not kill, and what shouldn't you kill? I ask this question because this 'commandment' is disobeyed and pardoned by humans frequently. To examine it, I will look at two types of laws: the first one is established, executed, and enforced by the Creator like, tomorrow will be another day; all planets must hang without physical supports; all living things must eat and sleep; everything will need time to become what it is created to be; each living thing will die after old age, etc. The Creator's law doesn't change. They are universal.

Human law is generally meant to protect, guide, and maintain the order for all the living things in the community that established the law. Human rules can vary from community to community. Some human laws may be for more than one community, state,

and even countries. One must drive on the right side of the road in the USA and many other countries. However, there are countries where one is required to drive on the left side of the road. It is acceptable to marry one's cousin in some communities but an abomination in others. These are human-established laws.

Yes, "Thou shalt not kill" is one of the ten commandments. It must not be God's Law, as commonly understood, considering the above analysis. As such, it is really no law at all. Let me give some examples to explain my point. When a plant is pulled from the roots to be used as a vegetable, the plant is killed. A tree is destroyed if it is cut down or uprooted for any reason – to make way for new construction, produce firewood, make papers, produce timber for shelter, etc. Then there are the animals!

Imagine how many chickens, goats, cows, different kinds of wild games, etc. killed every day for food and their hides. They number in millions. Some people engage in what they call 'sports'—shooting (killing) birds without thinking anything about it. According to God's law, as we learned from studying nature, it is certain animals need to be killed for food or when they are a danger to other animals. So, killing animals for food is not forbidden.

Please think of how many people are killed in our streets, roads every day: some accidental and many intentional, including domestic violence and drug wars. How many babies at different stages of life are aborted (killed) every day? How many people were killed in inter-tribal wars, international wars, and world wars? They number into the billions. Peoples, animals, and plants are killed every day, and many in the 'name of God' or for 'religious reasons,' like the inquisition, holy wars, etc. I submit that the statement should be rephrased as, "Thou shall kill only when necessary" – particularly for food and in self-defense (when attacked by deadly force). The "Thou shall not kill" will always leave the reader guessing what was meant, considering the number of killings witnessed or heard of every day.

If a man catches another man rapping his wife or daughter or even his mother, would he be found guilty for killing the molester if he could? No! This act will be termed 'necessary.' There are cases of killing that others view to be justified, such as abortion, when a rapist impregnates a woman, or a fetus is not viable or when the life of the mother/baby or both is in danger. Families involved in making these decisions will almost always choose to save the mother's life and not that of the fetus. Again, this is a necessary killing.

But if a pregnancy is due to uncontrolled indiscretion or the pregnant woman is poor, humans will be hard-pressed to allow termination of the fetus because the killing will be viewed as unnecessary.

Daily, meat, fish, vegetables are needed for food. Killings will be allowed to satisfy these necessary needs. These are essential killings permitted by the Creator since the acts are universal. The Creator gave humans the authority to make such decisions.

The Creator endowed humans with intellect and senses with which to discern and make decisions in both simple and complex situations. It is reasonable to believe that He also gave them the authority to make laws and impose consequences as they impartially see fit. If a person (even an animal) is a danger to society, the authorities must remove the person/animal from the community if treatment and or other measures fail. Removal may include death. From this analysis, it is easy to determine when killing in wars is necessary/unnecessary.

What of predatory animals like lions (even humans) that kill for food? So, it is designed by the Creator and, therefore, will not carry any consequences. If an animal goes on a killing spree, such an animal will be sorted out by the earth's caretakers and destroyed.

The same fate is suffered by humans that engage in the indiscriminate killing of animals, destruction of vegetation or crops, or killing of other human beings, like in terrorist acts. Such a person is sorted out and removed from the community and may be destroyed.

One might be interested to know why killing is allowed if it is necessary. From all discussed above, it is certain all creations – all living things – are essential and created for a reason, the most significant of which is the perpetual bond that is shared among all living things. None can exist in the absence of the other. All creations are therefore sacred, and their lives can't be taken lightly.

It is no wonder that in some cultures, before life is taken or any living thing is killed, the people will talk to the animal or plant to make peace, thanking the plant or animal for sacrificing its life for their sake. Some communities will never kill any animal under stress. Yet, many communities kill with no regard for life daily. Imagine the consequences of killing animals in that condition, not to talk of for sports.

Chapter 15

IT IS THE CREATOR'S WILL

The above statement is a popular saying among people, particularly those from third world countries. I have lived in America for many years and interacted with people there and from other advanced countries. The expression, "It is the Creator's Will," is rarely used among them.

There are numerous forms of that saying, like: "God knows the best; Everything depends on God's Will," etc. There are many events in the lives of these people when they feel desperate and helpless. At that point, they have reached the end of their ropes. Some of the events that can trigger that feeling could be the death of a child for whatever reason. The premature death of any person can evoke similar feelings. But dying at old age does not invoke such a state of helplessness. It is also heard when a person misses an opportunity, any opportunity.

'It is the Creator's Will'; 'God knows the Best' or any of the variations implies that the unfortunate event (or celebratory event) is according to the Will of God or that God sanctioned or authorized the event. Could this assumption be correct? Could God, the maker, and provider of all the good things of life, endorse the death of a child or allow an opportunity to be missed? Let's dig deeper.

When I was growing up, there were high incidences of infant mortality. But now, with more education on the subject, improved healthcare, and delivery, the situation has improved tremendously. Then, couples gave birth to as many children as possible. The rationale was that some would die, but likely a few will survive.

The mothers of the newborn babies suffered a similar fate: Getting pregnant or delivering a child was like going into the battlefield – you may or may not return alive. But now, it is almost sure the woman will live.

In the United States and other advanced countries, infant mortality or a woman dying during childbirth is nearly a thing of the past. People are educated in these matters. Besides, there are so many resources and procedures at the disposal of healthcare professionals. These resources make losing a baby and or the mother during child delivery almost improbable.

A good question here is: What is the relationship between a person dying or missing an opportunity and the Creator? Practically none! Granted, the Creator of all things, visible and invisible, can do as He pleases. But why would He randomly take away or destroy what He provided for the enjoyment of His creatures? Of course, we can't go to Him for the answer. Then we employ our usual – examine His creations for the answer.

He provided the earth with sunlight, air, water, food, plants and animals, flowers, the soil, etc., for the use and enjoyment of His creatures. Has there been a time when He took away or destroyed any of these blessings for any reason? Never! God does not take away or destroy what He created or provided. Therefore, the statement: "It is according to the Will of God" or such opinions cannot be correct. It is an absurdity.

Also, we need to understand that all the untimely deaths (what may be mistakenly regarded as 'taking away' by God), for example, are due to human activities, as explained below. However, after ripe old age, death is according to the Divine order: every living thing should reach its ripe old age for its kind and then die. How can this statement be verified to be true? I use the usual method—examine the creations of God, all living things.

From what we observe in our environments, we know that all living things started life as a singular cellular organism, which is called a 'seed' (more accurately, from 'nothing'). When fertilization (the seed enters the egg) occurs, the 'seed' turns into a seedling, if a plant or 'baby' if an animal (simple explanation?). Eventually, this 'baby' turns into an 'adult.' Then, the adult produces its seed, which will start a new cycle (see Perpetual Cycle fig. 6). This adult (plant or animal) continues the journey of life past ripe old age. Then death extinguishes the flame of life, marking the end of life.

Going through these stages is the order, the way the All-Knowing Creator established it. There is no other way. The fact that death after the ripe old age is not mourned but celebrated supports the authenticity of the order. It is easy then to understand why life that abruptly ends is mourned. Because we don't see it happen that way in nature death is mentioned in the perpetual cycle only at the end after the ripe old age. Therefore, it makes sense to believe that nobody, no living thing, should die but after old age for its kind. But we do witness deaths before ripe old age. Could this be an act of God, or is something else interfering with the Divine Order? Could anything interfere with the divine order? Consider the following:

The Creator gave free wills to all humans. God does not interfere with the free will He bestowed on humans. However, the Creator provided humans with all the resources to make choices, including the intellect, senses, etc. Therefore, humans have the responsibility to make use of the resources.

Could death be a consequence of disobedience or an act of God? We know the results if the divine order, "You have to eat, sleep," is disregarded. And we know the Creator can do as He pleases but does not take away His blessings to His creatures. Is the saying, "It is the Creator's will," then false?

Let's examine a few instances: A woman had a miscarriage of two months of pregnancy. Another young woman could not even conceive after many years of trying. Another scenario is a child was born with injuries on some parts of the body. There are instances where some babies were born with disfigured limbs. Yet, some babies are born with traces of cocaine in their veins. A woman lost her child because she was too poor to provide food for the child. A boy playing with the father's loaded pistol shot dead the friend.

A young man visited his usual bar and had three to four shots of whiskey and left the bar only after topping up with two bottles of his favorite beer. On his way home, he fell asleep behind the wheels. He died at the hospital. Four brothers joined the army together, but none came back alive after their first battle.

A man and all the members of his family were killed in a drug war. Two family members were killed in their own home by armed robbers after their brand-new car.

A teacher and four of her students were killed in a school shooting incident. A woman was shot dead by a jealous husband. One hundred fifty people died in a plane crash due to a mechanical fault.

Food scarcity or a hostile environment will kill many animals, including humans. If a forest is set on fire for whatever reason, all the inhabitants that could not escape will die.

A seed planted on top of stone or in muddy or sandy soil will eventually die. Yes! Even if the soil is good, the result will still be the same if the seed lacks other necessary needs for growth and survival. This list can go for miles, all the incidences occurring before the ripe old age.

Before "ripe old age"! No, or did somebody disobey the Creator's law or failed to use the gifts of the brains, senses, conscience, etc.?

A careful examination of these incidents will show that all the deaths were preventable. Should humans know this, or are humans not equipped with the brains, senses, etc., to understand that these events will be apparent if the natural conditions are not met. Or that the Creator does not interfere with our decisions since He has already given us the resource we needed?

Could the abortion of a fetus or finding traces of cocaine in the baby be an act of God or the consequence of past behavior of the parents like over-indulgence in drugs, alcohol, diseases, uncontrolled indiscretion, etc.? How healthy was the seed planted? Was God supposed to have sent one of the angels to put away the loaded gun or warn the boy's father not to be careless with such a lethal weapon? Why should it be taken as the "will" of God if a child dies of malnutrition?

Did God take away the sunlight, rain, farm crops, the fertile soil, or did the soil refuse to yield and grow crops? Should we see it as a failure on the part of God for not sending His angels to repair the faulty airplane to prevent the accident? The operators and owners of the plane were not given the resources or the ability to check the aircraft for flight worthiness?

Also, God was blamed for not preventing the brothers that joined the army 'to kill or be killed.' Was God supposed to interfere with their free wills? Just imagine some people standing akimbo watching as fire destroys houses and buildings, farmlands and crops, forests, and even with humans. At the end of the inferno, some people will declare, "God gives, and God has taken." Some people may wonder why God did not send His firefighters to put out the fire.

You can see such people failed to use their brains and senses to acknowledge the disastrous consequences of a fire event and made preparations in advance against the occurrence.

No! "That will happen if God permits it or if it is the will of God," some people will chant!

Humans foolishly blame God for these incidents because they forgot that the Creator is constant. He does not change His mind about anything and certainly not about interfering with the free will of humans?

It is not right to blame God when such a disaster occurs. Instead, we should look at and blame human errors, their inability to use God's gifts of the brain, senses, etc.—their disobedience.

There was a story about a man who built his house very close to a tree. His friends advised him to cut down the tree because it was too close to the home and would pose a danger under storms. The young man was more interested in the shade the tree provided and was encouraged by his firm belief that everything was in the hands of God. "My house will be protected," he assured himself.

Later, there was a wind that brought with it heavy rain. The obvious happened – the tree was uprooted and fell on and destroyed the house. The man should blame himself, not God – he did not use the Creator's gifts of intellect, senses, etc.

As stated earlier, there was a time of high infant mortality in my country. But now, the frequency has been reduced considerably. Is it possible that then God was sanctioning or authorizing the deaths of those children, but now He has decided to stop? That would not make sense! The Creator is the maker, giver, and sustainer of life. So, what happened?

Education and improved healthcare, sanitation, and nutrition! It is not the will of God to take away His blessings to His creatures. However, it is His will that humans use their gifts of the brains, senses, etc., to better their lives on earth.

These assumptions, shifting blames, and ignorance are at the core of a lack of upward mobility in many countries, particularly third-world countries. Yes, 'Leave everything to God, blame God.' It takes a lot of work to use the brains, senses, etc., the Creator gave us and free? It is also why these peoples swallow the baits of deceitful 'men of God' who convince them to pray and worship: all they ask for will be granted. Can any person show anything given because of worshipping and praying? Is everything in the hands of God in that sense? It is not valid!

It is the preacher and their organizations that profit most from all these. The 'man of God' understands the psychology or the thinking of the blind-faith worshipers: All he needs to do is 'touch some nerves,' stir up emotions like guilt, need, regret, etc. and then reference some 'Divine Promises' of everlasting bliss after death and watch his sticky fingers in the wallets of the gullible.

On the pulpit, the names of 'past holy' men and women are evoked, names like Moses, Jacob, the Apostles, etc. They have convinced their flocks to believe that any person that 'has faith' will do great 'things,' even 'greater' than what Jesus did. You wonder why they don't quote the names of some 'Holy' men and women from their groups, people still alive—Imams, preachers, reverends, nuns, ministers, bishops, powerful or less influential televangelists, etc.

Do these 'peoples of God' serve and follow Jesus and other prophets? They say they do! Then, they must have faith and must have been doing great things or doing things greater than Jesus did! Has anyone seen or heard of a 'man of God', a 'Holy' man, anywhere that has been doing great things or doing greater things than what Jesus did because of his faith? Is it possible that there is no such a person, no 'man of God' that has faith? If such a person does not exist, then the only plausible takeaway is: They don't even believe in what they preach and, as such, are false prophets, the equivalence of armed robbers but in white robes.

The naïve; those who mortgaged their God-given gifts of the brain, senses, the conscience, etc., to their 'holy people'; those whose ears are plugged, and eyes bandaged close; will continue to go to their 'holy leaders' for spiritual guidance, unfortunately. Hence, their holy leaders become wealthy from the donations of their ignorant 'faithful' followers. These leaders can't tell who the Creator is because they don't know. They are pretenders and deceivers.

You may also wonder why there are different translations or versions of the 'Holy' Books. Let's consider one of the holy books, the Bible since it is one of the most popular. There are nearly 2,000 versions or variations, while the King James version appears to be the most used.

A book can be translated into many languages. But the book remains in tack – no change in the message of the book. However, the users of the Orthodox translations can't think of using the King James version or other versions. Don' you wonder why?

The Creator and only the Creator is One. So also, is 'His Word'! Any holy book, the 'Word of God', must be one. That would mean that different versions or translations must be only for convenience or ease of understanding, not that they convey different messages.

It must be interchangeable! Since it is not so, does that not call to question the authenticity of this Holy Book? Googling the events at the Councils of Nicene and Chalcedon in the fourth and fifth centuries, respectively, that produced the book, answers that question.

Any law from the Creator for His earthly creatures is universal, needing no specific language, unique talent, or high position to translate or understand. It can't even be written down in words. Or in what language? In what language is the law to female lions (or any animal) on how to mate, bear and raise their cubs written? How does the Creator communicate to plants the time for growing their roots, leaves, fruits, etc.? None of these were written down. Instead, they are written in the natures of every living thing.

Animals and plants are earthly, while the Creator is a Spirit. He designed these things. The Creator knows more about plants and animals than they know about themselves. Yes, His laws are etched in the DNA of His animal and plant creations. So, any laws that require specialized knowledge or a particular position to translate or understand can't be of the Creator.

Parents can only direct, show the young ones the ways of life. These already exist in their DNA. But how they are developed and applied will depend on the influences in their environments. When a mother places a baby bottle of milk in the baby's mouth, it will suck and swallow. The baby did not need to attend classes to learn how to drink anything. What of the first time the baby stood up? Yes, the baby will stand when the time comes. What of when to cry, smile, laugh, eat, sleep, ease itself, etc.?

When a seed is planted, the farmer does not write out instructions to the grain on when to grow its roots, leaves, how tall for it to grow, how many fruits it should bear, etc. No! It is already written in its DNA.

Granted, humans need to write down things for different reasons. But the Word of God and its meaning can't depend on people's opinions, whether 'holy or ordained.' God is the writer and does not need a pencil and paper.

If knowing the Creator, our duties to ourselves, and all God's creations are already written in our DNA as in plants and other animals, what is the need for a religious leader or holy man; why the hate, evils, and unhappiness of this world? A simple answer: the outcomes of the influences in the environment – particularly the family, religious leaders, and the preachers.

Because of free Will, some men have declared themselves "holy," arrogated to themselves the power to forgive sins. These "people of God" have tricked their followers into trusting them and contributing their widow's mite to provide them all the luxuries of life and also to erect edifices, ostensibly to house and for the glory of the Creator. But they know that the Creator of all things is not bounded by space and time. He does not need or cannot be contained in any building or anything made by man. By brainwashing people to believe and trust them, these leaders were able to spread unhappiness to our world by teaching falsehood about the Creator, God.

Please do yourself a favor; read the Creator's laws as written in His creations around you, including using the free brains, senses, conscience, etc.

Chapter 16

DIVINE DETERRENTS

In the scheme of things, man is too little and too insignificant to alter the natural laws that God, the Creator, has put in place. To create living and non-living things, many people believe God first created the necessary materials! Can God use 'materials' in that sense? Of course, God can do whatever He chooses. But materials are earthly, and God, a Spirit, does not use or need material things. The Creator brought these materials into existence for our use.

Humans run into trouble and cause a lot of misery when they misuse or try to alter the composition of these materials, making them unnatural. We are witnesses to the disastrous effects of Climate Change due to the Green House Effect, a deliberate act of humans: like changing the natural composition of the soil that produces the food we eat by forcing it to have more; the genetic manipulation of seeds to make them resistant to different types of infectious diseases; the shots administered to cows and goats to make them fatter and produce more milk; the antibiotics and hormones fed to chickens to force them to live in crowded conditions and to have more; the abuse of our atmosphere through pollution and dumping of chemicals into our water supply; and so on.

The Creator gave humans the intellect and authority to use His creations to benefit our lives. He also gave each living thing adverse reactions to alien substances. Is it any wonder that much of the adversity we experience today is caused by the attempts of humans to interfere with or change the Creator's natural order?

Consider the advances in Science and Technology like GMOs (Genetically Manufactured Organisms), growing human tissues for repairs, growing animal flesh for food, artificial insemination, cloning animals, etc. In some quarters, these acts are unnatural and maybe against the intentions of the Creator. But are they?

God has designed all things and wants them to exist forever. These include the ability and intellect given to Humans to use and do as they see fit in caring for all living things and the earth. He does not want humans to change or try to redesign His creations. I explained elsewhere in this book how one can reach this conclusion by observing nature.

Before attempting a redesign of any object, one must first understand the workings of the thing and the designer's intentions in making it. Humans tried to redesign the food we eat without realizing that our body is made to eat and function with the type of food created for its use by God. We are already aware of the havoc unnatural foods cause in our bodies.

To be sure, He put in place deterrents and consequences to prevent those ever happening. The restraints include the imprint: Ns factor—the conscience or shame, guilt, etc.; human brains, conscience, etc. Every living thing has the mark of the Ns factor.

What is the Ns Factor? It is what makes every living thing act unconsciously or instinctively, generally with no apparent reason. A few examples will suffice: Place a bottle on the lips of a newborn baby; it begins to suck it. Bury a life seed, any seed, in the ground, particularly a fertile ground containing moisture in the presence of sunlight and air; the seed will germinate and knows to face the sun. If someone molests or maltreats any person in the presence of another person or group of people, someone among the spectators will start to fight against the injustice. These are a few examples of the Ns factor.

Consider the problem of global warming: While some people refuse even to acknowledge its existence, many are protesting and taking action against it. Another deterrent put in place to discourage altering any of His designs is the consequences of such an attempt. We know the results when any living organism refuses to eat, breathe, sleep, etc. Sickness is a symptom that something is very out of order.

These would amount to disobeying the orders of the designer who has commanded that all living things must eat. He knows that our bodies need oxygen and has designed the means for our system to extract it from the air we breathe in. And He made the night for rest, during which time the body will rejuvenate or repair itself and made the day for work.

There have been fears in some quarters that considering the level of sophistication in arms development, humankind will one day destroy itself and end the world as we know it.

The destructive powers of the atomic bombs dropped in Nagasaki and Hiroshima are now a child's play compared to the current versions. They are truly frightening!

How can the envisaged Doom's Day take place? Did they forget that only the Creator has the power to end the earth as we know it? Or did they forget the Divine deterrents in place to make sure that never happens? We know the efforts being made by both nuclear and non-nuclear nations to end nuclearization. Many countries have or want to have it only to have a seat among the community of powerful nations. There will never be a nuclear war that would destroy the earth. Only the Maker can destroy it. However, the fear that this invokes gives tyrants and powerful nations power over others.

The humans' dependency on one another that was intentionally designed by the ever-knowing Creator, alone, will discourage Nuclear War from ever happening. With the development of the Internet, the world is closer to becoming what is now known as 'Global Village.' News or events from any corner of the earth are transmitted worldwide in a matter of seconds. Disaster in any part of the Globe is no more country-specific but the World's problem. We know, for example, the work of Doctors Without Borders. The help of this organization comes in the form of money, equipment, skilled services, etc., that are rushed to places when disaster strikes.

Remember the Valdez oil spill, where resources of the world were assembled to save the environment. Even birds and fishes did not escape notice: Birds were washed clean one by one and released back to the wild, their natural homes. Will the rest of the world standby and watch a more powerful country annihilate a less powerful one? No! Then how can a person, a group of persons, or even a powerful country ever think they can destroy the earth?

It has been identified that the origin of most of the unacceptable behaviors that cause strife here on this earth has its roots in the indoctrinations and experiences that mold a child. In the beginning, the minds of children are clean and innocent, like a clean slate.

We know what the accumulation of writings and erasures do to the clean slate over time. Similarly, the messages recorded in the mind of a child result in the behavior or personality of the future adult.

Included in these is another Divine deterrent – the attitude or the rebellion of the young peoples of the world. The behavior of these young people is the hope of the earth. It is the youth that has risen in opposition to their parents' indoctrinations, injustices, and discriminations of all types. They said, "No!" to global warming, wars, use of chemicals to alter the Creator's designs. They do these by organizing rallies like earth's day, Global Citizenship Day, active involvement in the day's politics. They are less judgmental and teach and practice the love of one another as they see themselves. The availability of social media is a blessing in this effort.

They have come to understand the power of love and embrace their attributes. Selfishness (the origin of all evils) is sent packing where there is love. They realize that no evil can stand against love, and with love, everything is possible. As the Creator intended, the earth will ever be the home of all living things.

But humans do many things out of convenience and necessity. For example, a baby is supposed to be delivered through the birth canal. Considering the pains the mother must endure,

humans may be tempted to question His wisdom in prescribing this method. But the Creator is All-knowing and understands the necessity. Recently, Scientists understood that natural birth is best for the child and mother.

If, for any reason, a woman can't have a natural delivery and a forced attempt will kill her or the baby or both, He has equipped humans with the brains, the sense, etc., to deal with such situations. So, using C-section to deliver the baby will save the baby and mother.

But in some instances, women use C-sections for convenience; they don't want to be disfigured, they don't like the inconveniences and pains of natural childbirth. The unnecessary use of C-section would be disobeying the Creator's law, which carries consequences.

Chapter 17

THE EVERLASTING DIVINE BOND

Nobody can see or talk to the Creator. We can only try to understand His intentions by examining expressions of Himself – His creations. There is nothing in His designs that can give us a clue for the real reason He created human beings. Humans can understand God's intentions for providing foods and making them available forever. It is easy to see His intentions that human beings and other living things will be on this earth forever. The world is the home of humans, animals, and plants. We know these to be true when we consider that our planet is the only known place where animals' food is provided. The foods are tied together in a "one for all and all for one" everlasting bond.

None can exist in the absence of the other. A few examples will suffice: Animals breathe in air, extracting the oxygen, in part, to oxygenate the blood. One of the byproducts of this process is carbon dioxide. Without carbon dioxide, there will be no food for plants (photosynthesis). Plants will die without carbon dioxide. When this happens, food and medicine for animals from the plants and vegetables will disappear, as will the animals and humans. The sun provides warmth to all living things and rejuvenates the earth; photosynthesis can only take place under sunlight with a byproduct of oxygen, essential in the existence of animals. So, all plants will die without sunlight, and animals that depend on plants for food and medicine will disappear. If there is no soil, particularly fertile soil, crops and plants cannot survive. There go animals and plants. Furthermore, if there is no water, plants and animals will die since all living things require water for different functions in their bodies. All these materials are abundantly and freely provided.

What is the lesson from 'everlasting bond?' That all the living things will exist forever 'provided' that the natural order is maintained. We know the natural order must be held because of the Divine deterrents and consequences He put in place.

Let us examine water or moisture further. First, it has been established that the Creator's intention was for all living things to exist here on earth forever. We know this from the fact that He provided the means (the 'seeds') to reproduce themselves and keep the cycle in perpetual motion. Second, there is nothing in the physical world that indicates that the oceans, seas, rivers, underground water systems, etc., of the world, will ever dry up. Besides, with the Maker's recycling process, the sun's heat induces evaporation, which later condenses and falls back to the earth in the form of precipitation.

The resulting runoff performs its critical functions, beginning with washing and cooling the atmosphere, watering our vegetation, hydrating living things, replenishing the water, oceans, seas, rivers, lakes, and streams lost to evaporation and what remains joins the underground water systems.

It is a fact that through procreation, humans and other animals, including other living things, will perpetuate their existence here on earth. Besides making the earth 'ready first' before creating humans, He gave humans the brain, the seat of knowledge, and planted in them and all living things the Ns factor. The brain is the most complex of all the systems of the human body. Scientists have shown this to be true.

One can lose a leg or both, lose an eye or both, a kidney, lung, and still live. But not the brain! Human knowledge of the human brain is minimal, less than half of one percent. Other animals and plants have a primitive form of the brain, referred to as 'instinct.'

The instinct in no way compares to what He gave to humans: The brain – the intelligence— makes humans one of the weakest of all animals, superior to other animals. With the brain, humans can think, invent, solve complex problems, and improve their lot. But other animals can't do any of these.

The gift of the brain leads to the following question: "Why did God create all living things but gave only to humans this unique brain, the seat of wisdom, the conscience, and the right to choose? One is tempted to explain that He did so to enable them to take care of His earth, including everything on and in it! These included studying and understanding all creations, making laws, establishing rules and consequences, and ruling over the earth. The effects of using the brain can be seen in all communities, now grown into countries, that each has its own set of laws with their judges to administer and enforce them. You may see the same type of laws in all countries, while others are country-specific, depending on the people and culture of that country.

But in all these, humans failed to understand who the Creator really is. Hence, humans feel free to act and do as they please.

Indeed, humans don't know who the Creator is. Very unfortunate! It is almost inhuman or unthinkable even to imagine that the humans ignore the Being that created them and provided all their needs: the food, water, the sun, all the components of the air, the soil, all the beautiful flowers, plants, vegetables, all that makes life enjoyable and the life itself.

If humans have some idea about who He really is, how could they hurt any of His creations: How could humans kill unnecessarily, how could they be unkind, hurt, say or do evil things or carry out any form of wickedness on His creations? Please, reread my analogy – my friend on a telephone conversation with his boss holidaying in London.

This ignorance (No! The brains, senses, conscience, etc., are provided) brought into our otherwise beautiful earth all the evils and sufferings we witness daily. This lapse started at the beginning of human existence here on earth. Humans must go back to nature to understand who the Creator is and do His will!

Following the above discussion, it is clear the Creator intends to have all things exist here on earth forever through the divine recycling system, procreation, and the checks and balances He put in place. So, it is reasonable to conclude that the creations as we know them will end only if He wills it.

Also, one may conclude that the All-knowing God, the Creator of all things, visible and invisible, can take care of the earth without the need for human beings. Therefore, only the Creator Himself knows or can tell why He created humans and endowed them with the gifts of the brains, senses, the will, etc. The Creator is God and does as He pleases.

Chapter 18

UNCONTROLLED INDISCRETIONS

Some people (many times respectable ones) get or find themselves involved in unacceptable acts to society. Of course, the earth's caretakers are empowered by the Creator to establish laws governing human behaviors. Suppose a person turns into a swindler, rapist, child molester including pornography, human trafficker, killer, or declares himself an enemy of the society; how would the authorities deal with such a situation? The law enforcement agencies would use established laws to control or eliminate the threat and reduce the heartbreaks that any of these acts can cause to society. Sexual misconduct is dealt with similarly.

We all know that the Creator of all things is good, and His creations are also good. Societies do not turn their back on these misguided people but employ different strategies, including education and rehabilitation, to deal with them. More extreme measures could involve imprisonment or even death.

Many men of power and means have been pulled down from their high offices throughout history because of sexual indiscretions. Recently we witnessed the disgrace of many Catholic priests and bishops because of the same offense.

We wonder how these behaviors infested our beautiful earth, that the Creator generously furnished with all that would make humans live here forever and happily. Previously we identified that events or the environments in and or outside the family to be the culprits.

This section of this book examines the sexual behaviors of highly placed individuals or any person and tries to understand whether humans failed to read the Creator's laws as written in His creations that led to the misunderstanding of how to handle sexual morality? Hence, this book examines nature, the very handiwork of God, to understand the laws guiding sexual behaviors.

From the analysis conducted earlier, the Creator's intention certainly is to have all living things exist here happily forever. To achieve this, He made all the necessary arrangements, among which was procreation. All living things must do this, no exceptions, though executed in various ways depending on the nature of the living thing. Generally, all-female living things go through the 'flowering' stage, during which time fertilization, through a compatible male, may occur. These details though scanty, are all that is necessary here to explain how living things will maintain permanent existence here on earth.

In plants, there are no restrictions or control as to flowering, fertilization, or seed production. Each plant will flower when and as the Creator designed. It is almost unthinkable for a farmer to complain of too many seeds or fruits coming from his plants. He can't tell a plant, such as an orange tree, to produce only a certain number of fruits. The same thing applies to other animals like a dog. No one can tell a dog to have only one puppy or ten. These living things obey the Creator's prescriptions to the letter.

With humans, the case is not and should not be different but controlled by Divine Laws. In the design of human biology, nothing shows that the Creator intended to impose any form of restrictions like He didn't do with other living things. However, He gave only humans intellect, senses, including conscience. And we know why. Therefore, all animals, including humans, theoretically should reproduce without constraints. Does it mean then that a man should have sexual relations with his sister? What of the first, second, third, fourth, or even fifth cousin? Are there cultures that support men having such a relationship with their sisters, cousins, or even their mothers? As stated earlier, it is legal and even praised in some countries for men to marry their cousins, while in others, it's taboo. Let's examine this further.

Making babies (breeding) with a close blood relative, referred to as 'inbreeding or incest,' was tried in the past. It was found that the closer the blood relationship between the male and female, the more likely the offspring are to have one or more defects or congenital deformities. This outcome is factual and does not need a laboratory experiment to prove it. It is universal. This unacceptable result serves as a divine deterrent. A man's mother, daughter, or sister is out of the question because the blood relationship is very close. There are fewer problems with cousins, as seen in some countries. But as a rule, the more distance in the relation, the better. It is a human-made law because it is not universal.

You don't see Billboard advertisements prohibiting sexual intercourse with a close relative. Within some family circles, they could practice incest without attracting any attention. But why does this rarely happen? Because the control is built into the Creator's system. People might practice this in secret, but how can they hide the disastrous and shameful outcomes? It is incredible; children learn early in life that the practice is a no-no; they can't get sexually involved with their sisters or close relations. And in most cases, the command is obeyed to the letter requiring no supervision, proving that the Creator's hand is in it!

Unregulated 'procreation' can result in overpopulation or overproduction, threatening the existence of the living. As always, overproduction will trigger control or deterrent reactions from the earth's caretakers. Overproduction of food items does not pose as much of a problem when there are always the hungry to take advantage of the surplus. Of course, some people don't view the earth's bounties as the Creator's but their own. Such individuals would prefer to sink the excess in the seas rather than extend help to the needy to keep control of their bottom line.

Overpopulation of certain animals can be a threat to other animals and tilt the scale of the ecosystem. If there is an overpopulation of lions, tigers, certain killer birds or rabbits, or some types of insects, roaming space and the food supply will be contentious. Animals will kill their kind and other animal kinds for food. There will be a struggle for freedom with some animals straying into others' areas.

The Creator has already empowered the earth's caretakers to deal with the overpopulation in such events as they see fit. There was once a swarm of locusts that descended on some acres of cornfield. The problem with this type of infestation is the suddenness of the attack. No warning! Before the farmer knew what was happening, the insects had devoured the whole cornfield. As they ate, so did they multiply in number?

The farmer did not idle by and watch his farm and crops being destroyed. No! He employed different means to get rid of them. Though, at the end of it all, nothing remained of the farm or crops except heaps of dead locusts, protein food for the humans, and other animals. The earth's caretakers are empowered and equipped to deal with this type of event.

With human overpopulation, the game is different. No laws can support exterminating the excess. From all indications, the Creator also does not sanction extermination. The earth's caretakers designed various means to deal with this when it came to human overpopulation. With their populations in billions, India and China have experienced the dangerous consequences of not controlling human overpopulation. Uncontrolled population explosion will stifle the upward mobility of many children in the families, cause hunger and reduced quality of life. Children from such families will not live up to their full potentials.

Because of ignorance and poverty, many people will continue to make babies, claiming it is the Creator's gift, and He supports having as many children as possible. Not too fast! He gave every person the brains, the senses to be able to manage themselves. Who will provide for the children of these parents when they cannot?

The governments of China and India started attacking the over-population of humans with mass education to increase knowledge of what was at stake and discourage having too many children.

How many children are too many, you might ask? Only the family pocketbook can answer that question. The family bank account will dictate whether the children can receive a good education, have an acceptable quality of life, and be a burden to society. As a last resort, China instituted a law prescribing the number of children a family was allowed to have.

Coming back to indiscretions, what role does sexual impropriety play in all overpopulation? Really none! It is the moral aspect of the behavior that is under scrutiny here. But the Creator's laws are in place already. Yes! It is clear to every man never to have sexual relations with his mother, sister, or daughter. This practice was learned as a child. The same thing applies to any woman with a close blood relationship with the man. This ban would include nieces and aunts. In some countries, people are allowed to marry their cousins, as observed earlier. Such practice is an exception, not the rule. It is human-made, not universal.

In all the countries of the world, no man is allowed to have sexual relations with a child – a young girl who has not reached 'adulthood.' It would be seen as child molestation or rape and carries a heavy penalty as a deterrent.

It is debatable what is the definition of adulthood. Many societies or cultures have determined that a sixteen to eighteen-year-old female is a woman and matured enough to carry babies with little or no medical or health problems. In modern societies, girls need to complete high school or even college education before thinking of marriage. By the time this is achieved, adulthood is reached.

There is a dilemma in many advanced societies regarding dealing with the outcomes of 'maturity.' Because of 'improved' lifestyle, agricultural and food additives (such as hormones), sexually stimulating public messages, drugs like birth control, it is a common sight to see a twelve-year-old child (or less) a fully-developed woman with all adult features acquired and in full display. Such children can easily be mistaken as adults. In those societies, there are children, particularly females, that are 'sexually active.' The result modern lifestyle is that at a very young age, like ten, these children have started having sexual desires and involved in sexual activities. No divine or societal laws prescribe the onset of sexual activities, particularly before emotional maturity is reached. This situation calls for parental (mainly the mother's) moral instructions and coaxing to come in. It brings a lot of shame and dishonor to many families if an underage female is seen behaving sexually with men or involved in some nefarious activities like drugs and drug dealings, alcohol, etc. The results of such behaviors could be child molestation, child trafficking, slavery, diseases, pregnancy, and even death. Abortions from such behaviors can have a lifelong impact. Boys are also involved in these behaviors. But girls face more challenging consequences.

Some families have taken a variety of measures (many times, extreme) to put this type of behavior in check. For example, in some cultures, female circumcision, early marriage, early initiation of birth control, and involvement of religious leaders may be employed. In this age of freedom, selfishness, and disrespect, parents are left almost helpless. If a young person can't be restrained by parental moral instructions, family models, and love, sometimes physical force, soliciting help from even extended family members, many families give up.

It has been established that the Creator forbade a man from having sexual relations with the mother, daughter, sister, or any relative with close blood relation (no matter the age). We know this to be true because this law prevails without supervision throughout the world.

It is universal. Men with no close blood relationship can have sexual relations with these females only in marriage or by mutual agreement provided the female is of adult age and 'unattached.' Any other reason will result in a case of rape, which will incur a stiff penalty.

Also, no person can have sexual contact with another person's wife or husband. As will be explained later, this rarely will be a case of rape since it is an agreement between two adults without a close blood relationship. Yet when this happens, the consequences could be deadly (a deterrent). Why?

Suppose two individuals contracted the act with no close blood relationship and under their free will? We know that there is nothing in nature that restricts the two adults regarding sexual relations.

Then what is the problem, or why should the consequence be deadly? It is like stealing, betrayal in the highest order! No animal can dare to take/steal another's young one's food, usurp its space, and go scot-free. If this happens to a man, the result will always be bloody because his whole manhood is called to question. Every animal, particularly males, must defend and protect what belongs to them. Based on free will, a wife or husband was established by marriage to belong to each other. The deadly reaction by the husband/wife is a natural deterrent. It is universal.

One other scenario deserves mentioning here. Here is a woman, an employee, and her boss. They traveled to a business meeting. In between the meetings, the boss and employee took some time off to relax. It is during this time the boss has a 'sexual affair' with the woman. There are many reasons why a woman might yield to the man with little or no resistance: How could she 'disrespect' a boss who has her future salary, employment, and promotions in his hands?

In many countries, this act is frowned at as a part of the business, and the boss goes free. But in countries like the United States, if reported, it can cause a lot of unhappiness, as evidenced by the fate of many men of power and some 'men of God' recently.

From this analysis, sexual relationships are forbidden but allowed only when under an agreement by two adults with no close blood relationship and does not 'belong' to another person, particularly in marriage. Excluding these restricted areas, nothing in nature shows that a sexual relationship between two uncommitted, consenting adults is wrong or a 'sin' against God or society. It does not require rocket science to understand that males and females are meant for each other to satisfy the Creator's purpose – for living things to be here on earth forever through procreation. They are required to enjoy themselves here, on earth. The Creator provided all their needs, including all the things that satisfy the senses, to facilitate a happy stay. Just check out their biology: they fit together perfectly. So, naturally and in normal circumstances, opposite sexes must cohabitate or have a mutual need for each other.

But when I was growing up, I was taught that having sex (except in marriage) was a sin against God. Is this act really a 'sin'? There is nothing in nature to show that it was not the intention of the Maker for mature, available males and females to meet, enjoy each other and eventually procreate together.

As stated earlier, there are deterrents in place to discourage or prevent unacceptable intercourses. Is intercourse a sin against God, or is it an act that should be discouraged, particularly among young people, simply because of the inherent problems associated with the action: problems like pregnancy by a young person that is physically, emotionally, and financially unprepared; diseases; family shame and a host of other issues that are possible?

Considering what happens in nature, the latter reason applies. Sex in that context is no sin, but a possible responsibility and devastating consequence must be considered before engaging in the act.

It is obvious the Creator encourages it and prepares young people for it. The sex drive is so strong that it can overpower reason and even parental authority if care is not taken. All adults, particularly men, can remember what happened to their 'manhood' early in the mornings when they reached puberty, the number of times per hour they thought about sex, or how their whole body felt when they thought about their girlfriends.

I am sure women had similar experiences in their young days. We know how a man or woman feels when denied sex by a partner! We understand how a person feels when there was enjoyable sex the previous night or what happens to couples on their wedding night. These feelings, behaviors are natural, so wired by the Maker. It could not be a sin.

Before puberty (until recently), boys and girls rarely have sexual desires for one another. You may even hear a boy saying, "I don't like girls, or I don't want to play with her; she is a girl." Wait until sometime after puberty. The unquenchable fire of sex will start. Nobody teaches a young person the time to desire the opposite sex – it comes naturally.

A few years back, the young boy who did not like girls now spends more time thinking about girls or involved in foolish stunts to attract girls' attention. In no time, 'passing' notes in classrooms, phone calls (these days), requesting a friend to speak on his behalf to a girl, complimenting, openly admiring, and sharing little things with the girl of his dream begins. Before you know it, dating will start. These behaviors are generally restricted to males until recently. Why? Males are naturally the 'hunters'—the way the Creator made it.

But in these days of 'freedom' and 'equality,' especially in Western Countries, particularly America, the old order is fast changing. Complimenting, admiring, touching, and even looking could get somebody into great trouble. You dare not kiss or force yourself on a person without permission. These acts now go by different names like evading one's privacy or violating one's human rights, sexual assault, taking undue advantage due to the pursuer's power or position, or outright sexual molestation.

When it was acceptable to tell an unattached woman that she was beautiful, I still remember that her clothes fitted her perfectly well. Don't get me wrong, some women still appreciate those compliments, including opening a door for her, taking her out to restaurants, movies, etc. But how can a man be sure of who does or doesn't want to be treated that way in this age of "What a man can do a woman can do"? One looks before leaping or does so at one's own risk! It is worst in work environments. If one is not careful, the woman concerned will report him to the authorities for making sexual advances in a casual harmless joke.

It is not true that a woman can do everything a man can do or vice versa. Can a man get pregnant or carry a baby in his 'womb' for nine months or breastfeed a child? Can a woman impregnate a man? These questions should not even be asked. Why? They introduce an unnatural element of competition! There is nothing in nature to suggest that the Creator intended that females can or should compete with their opposites, particularly in a family setting. Every creation of God is unique, specially designed for a definite purpose, and must play its parts for the good of all living things.

Men and women, like other animals, must produce and nurture the next generation of earth caretakers and need to be happy in doing so. The Creator so ordains it. In this context, should touching, kissing, sex, statements like: 'I love you,' 'you are beautiful,' etc. in human interaction be discouraged? No! Doing any of these is how relationships, love, marriage, etc., begins. There is nothing wrong with any of these behaviors.

Something must be amiss if men of power and wealth or anybody can be pulled down from their Ivory Towers due to behaviors associated with sex, a prescription by the Creator. It's obvious the method employed to achieve them is the culprit.

Eating is required for the good of our bodies. It is dangerous to be eating while running. Sex under normal situations is good, as explained earlier. But bad or condemned if one imposes himself or uses force on the victim, i.e., it was not an agreement between two mature, available adults. Sex is for procreation, pleasure, or both and between two grown adults under the mutual understanding—the overarching conditions.

These days we hear of men jailed because they 'raped' their wives. "Raped their wives"! Yes, you read it correctly! Here are all the conditions – they are adults; no close blood relationship; the act is either for pleasure or procreation or both. But, the essential requirement, 'mutual agreement,' is missing.

It is easy to understand why consent is essential in this act: a man or a woman comes back from work in a rotten mood because of the events at the workplace. The partner wants to jump into bed, careless or non-sympathetic to the mood of his/her mate. This could result in an unpleasant end. The outcome is worse when a person uses his power of position or physical strength to have his or her way.

Some people employ deceitful ways to get sex – drugging a person or in a doctor's office when a patient is under anesthesia or luring the victim into a helpless situation where sex is forced.

In all these instances, what is lacking is mutual agreement resulting in the feeling of being violated. The feeling of violation should not happen to this act. Having sex is good. From observations alone, we know that the Creator sanctioned it.

Many have shied away from the discussion of what goes on behind a 'locked' bedroom door. One opinion is that what happens there is too private; it needs no scrutiny. But is that true? No! Because it is the place where everything about our existence here on earth started!

Check out the following scenarios: a little girl dashes into the daddy's bedroom while he was dressing up for work. Out of curiosity, she asks the father, "What is that sticking out in-between your legs; do I have one?" Or a little boy is asking his pregnant mother how his baby sister will come out from the mother's "stomach." Or a grown-up boy who wondered aloud to his father to explain to him why he had the slippery, colorless liquid that came out of his penis almost every morning.

Consider the general reactions of some ignorant and, many times, very religious parents to those innocent, natural, and childish questions. "How can your mind wander into such things? Are you going to be promiscuous or way-ward"? Thus, natural curiosity is snuffed out by ignorance.

Parents should explain to their children, young or older, such natural phenomena without shaming or embarrassing them. Just be honest and sincere, remembering that it is a part of parenting as expected by the Creator. Two lovers enjoy each other to the best of their knowledge. Nobody conducts lessons on how to make love. The knowledge is relatively intuitive. Young men and women also get some pointers from older people. The essential ingredient in this act is mutual agreement. If it is understood and adhered to the divine obligation to care for and love one another, nothing can go wrong in that relationship, no matter where the action took place.

Chapter 19

ARE WE TAKEN FOR A RIDE BY CIVILIZATION

Think for a moment about the progress humans have made in the march of civilization.

Compare it to where the 'cave people' were. Science and technology have been at the forefront in that march. Our fore-parents were naked and wandered about in search of food. What type of food did they eat? They ate wild fruits, wild games, and vegetables and primarily raw and in what we will now describe as an unhygienic environment.

Today, most of our foods come packaged. We don't have to wander about looking for them. Just walk into any grocery store or an open marketplace, and the choice is yours.

Then, food was consumed as soon as it was gathered. But now food can be kept preserved for months if not for years with canning, preservatives or stored in big refrigerators. Even scientists have discovered how to force a piece of land to produce food using different chemicals and insecticides. In the field of Cryogenics, seeds, animal, and human bodies can be preserved for centuries.

Advances in the production of meats, including eggs and fish, are breath-taking. With different types of feeds, primarily hormonal, a chick can be made ready for eating in less than two months. In the distant past, this feat took months or even years to achieve because chicks grew naturally. Then, eggs were small, and some chicks could hardly lay the regular one egg per day. Now chicks can lay more than one egg per day with sizes varying from small to jumbo.

What of cows, goats, etc., and their milk? Science and technology have discovered ways to manipulate nature to fatten animals and produce milk in more abundance. Many people drink more milk than water because of its availability, quantity, and low price. We will not forget the fish.

With a fish trawler, one person can empty a whole river of its fishes. Compare that method to when people used fishhooks or even bare hands to fish. How many fishes could bare hands catch? Just enough to feed their families. Now, the oceans and seas can no more supply sufficient quantities to satisfy our needs for fish. Consequently, people have turned to fish farming. With growth hormones, it is possible to produce as many fishes as needed in as short as possible.

Just imagine what the ancient people faced in the blazing sun, winter, storms, and wildfires. At best, in such hazardous situations, they would crawl into their caves and huts, hoping for the best. Now, many people live in modern homes and work in modern buildings, equipped with the best furnishings and equipment that science and technology can offer – clean and sanitary, air-conditioned with electricity giving them uninterrupted light. Natural sunlight is no more necessary. People can be awake for hours, work as many hours as they wish, or simply sit around to play or enjoy themselves. The electric light is available in many countries now. Even warfare has seen a lot of improvements. In the olden days, wars between tribes could, in the end, record a small number of deaths or no deaths at all. What were the instruments of war then? Hand-combat, bows and arrows, catapults, sticks, etc. But now, science and technology have produced frightening implements of war. We have different types of guns with various capacities and accuracies and can fire so many rounds per minute.

Included in this arsenal of war are the guided and unguided missiles and the drones. We also have different capacity bombs like carpet bombs or bombs that can penetrate any structure and destroy or kill anything in its path. With these, one does not need to be at the theater of war to kill.

With these instruments of destruction, a day's battle can record thousands of deaths. Finally, the ultimate war weapon: nuclear bombs! One atomic bomb can wipe out a whole city in a matter of seconds, and the by-products – radiation—can put an entire country in jeopardy for a long time. The effects of such bombs dropped in Hiroshima and Nagasaki toward the end of World War II continue to wreak havoc in those regions up to today. When these two are compared to what we have now, the differences will send shivers to your bones.

Advances in communications and travels are mind-boggling. I remember when it took more than a month for a piece of mail to travel from Nigeria to Britain. Very long-distance journeys were by ships or trains. Traveling from Nigeria to Britain took weeks, then.

News and events recorded the same speeds for transmission. But now! Anything taking place in Washington or any part of the globe will not only be heard but witnessed live in real-time. Many TV stations are open twenty-four hours. With the little hand-held devices like cell phones, you can communicate with anybody anywhere on this globe.

With the aid of space-based satellites, weather stations can predict current and future climatic conditions. With the GPS, the need for and use of maps are very limited and soon will become a museum piece.

In medicine, advances are incredible. Shots, vaccines, and drugs have put diseases in check. Infant mortality is almost zero in some countries. Now mothers don't need to go through the pains of childbirth as some women of means can opt for cesarean section at their convenience or have a surrogate inseminated to go through it all for them. Mothers dying from childbirth are now reduced considerably.

With the x-ray, mammogram, MRI, sonography, EKG, etc., diseases can be identified, diagnosed, and treated. Before TB, malaria, leprosy, polio, cancer, AIDS, etc., used to bedevil humans. Most of these diseases are now controllable. Healthcare and its delivery are now superb. As a result, there is a higher for older peoples. With improvements in mobility, home care, and security, nursing aides, and retirement income (Social Security, etc.), older people live independent and more fulfilled lives in many countries.

Please check out all the improvements in science and technology. Undoubtedly, your only conclusion would be that many things have changed supposedly for the better from how they were with the prehistoric people. Then you would think the inheritors of the fruits of science and technology would be happier, making it possible for all peoples to happily comply with the Creator's laws and fulfill the purpose for which they exist here on earth. Above all, peace and happiness should be raining throughout our earth.

Unfortunately, the reverse is the case as evidenced by the number of deaths we witness daily—accidental or self-inflicted; terrorism; wars and threats of wars; cyber espionage; selfishness and win-at-all-costs mentality; the mention of the word "God," the name of the Creator of all things in the english language, is forbidden in some quarters. Instability and divorce in families have given rise to family breakdown. These have led to fears, poverty, and unhappiness all over our planet earth.

A good analogy is like being hungry in the house of plenty or when your refrigerator is filled with all types of free food. Yes, the Creator provided humans with the brains, senses, and limited free wills to 'boldly go,' borrowing words from the Star Trek, 'to places where no man has ever gone.' He allows and supports it only if it enables humans to fulfill the purpose for which He put them here on earth.

It should be noted that the Creator did not ask humans to redo or recheck His handwork. Humans were not allowed to redesign or alter the Creator's work. Why? Because the Creator's work is perfect, and humans are incapable of making changes anyway. Any attempt to change any of God's creations incurs retribution – unhappiness instead. Everything that God created has been given its purpose from which it can't deviate.

Humans have foolishly and ignorantly tried to redesign some of the Creator's perfect work, incurring retribution for such attempts. Humans thought they could do better than God! Few examples of such foolish attempts and the disastrous consequences will suffice: When God created arable lands, He put in them all that crops and plants would need to grow healthy and fruitful. Also, God had put the food materials, fuel, medicine, and means of protection for all the animals in the crops and plants. He knew what the journey humans and animals were set out to undergo here on earth entailed. So He made it that animals, including fishes, also provide food, protection, services to human beings.

Some wild animals like lions and sharks sometimes make their food of humans. It is easy to see; it was not meant to be so. No animal goes about hunting humans for food. Such wild carnivorous beasts hunt for flesh. When they are hungry, any meat will do. On further consideration, it is easy to see that the ultimate intention of the Creator is to provide all that humans need to fulfill their duties here on earth. In other words, every other thing is created for the sake of humans.

God created and blessed the arable lands and commanded them to supply all that the plants would need to keep themselves alive and provide what all animals would need to live and fulfill their obligations. The Creator's design was perfect. But man, due to mainly greed, decided to alter the composition of the soil to force it to yield more. By using chemicals, humans were able to manipulate the soil to yield more than it was designed to produce.

The truth is that if the soil is used the way the Maker designed it, it will continue to produce food. That would be too slow for the need of modern humans! The use of pesticides on farm products like fruits, seeds, vegetables has rendered these products harmful to animals and humans when consumed. Dairy products, including chickens, eggs, fish, and meats, have lost most of their natural food values due to what they have been fed. Typically, chickens take about a year and a half to mature for consumption. Now it requires only a few months.

Chicken meats do not smell or taste like chicken anymore. In some cases, you might think you are eating tasteless cardboard. In some poultry farms, chickens can lay more than one egg per day—not as the Creator designed. Because of what cows and goats are fed, their milk lost most of its food value. Milk, whether human or animal, is designed to be a complete food. As God created them, these food items, including all farm products, meats, milk, and milk byproduct, serve as food and medicines. Modern science has robbed them most of these qualities. For example, see the sweeteners and sodas mentioned elsewhere in this book.

In the last few decades, science has started to genetically alter the natural composition of some seeds like corn, rice, beans, soy, etc. Note: "alter!" Humans are forbidden from changing the Creator's designs without incurring retribution!

Before I continue, please consider the following: When Scientists designed and built the car, they figured out and established that the only source of power to their invention was through the controlled explosion of gas/fuel/petrol inside it. Think of what will happen if, instead, we try to use water, diesel oil, kerosene oil, liquid hydrogen, or oxygen gases, etc. None of these will work simply because they were not in the original concept of that type of car. Some cars use diesel oil. But this type cannot use petrol or water as a source of power. Someday, cars that use water as a source of energy may be invented.

When the All-Mighty God, the Creator, fashioned the human and animal bodies, He also created the type of food suitable for His creation, the kind of food, when manipulated (digested), by the body as designed, would supply the nutrients and medicine the body would need to function. You can see why any other type of food would spell disaster in human and animal bodies. I will quote from an excerpt from the Director of the Laissez Faire Club, Doug Hill: "A little over twenty years ago, scientists started tinkering with our food supply, taking genes from one species and forcing it into the DNA of other species, creating new organisms that did not exist in nature . . ." He continued, "I'm talking about crops that have been genetically engineered to either produce their pesticides or withstand mega doses of herbicides." These 'new' crops are called GMOs or Genetically Modified Organisms.

Yes, the aim here was, among others, to produce perfect crops and seeds, able to resist infections, remain unblemished, look beautiful, last longer, particularly those that need to travel long distances to reach other markets. Ostensibly, the designers of GMOs were thinking of increasing food production to feed a growing population! Really? This statement implies that humans know better than the Creator; humans, not the Creator, know that the human population will increase and need more food to feed it!

The thinking is absurd and as senseless as a carpenter feeling; he could set a bar exam for prospective lawyers simply because a lawyer worked with him to win his case in a law court. No! There is nothing in nature to show that the Creator, the All-knowing, the Provider, left it for humans to increase food supply through altering His designs. Remember: "any attempt to modify the Creator's creation will only incur retribution." Is this not what we receive for attempting to change God's seeds, soils, animals, and plants

– inflammation, diseases, death?

Let's dig deeper into the disastrous effects of GMO foods on humans and animals. Now, most of our seeds and legumes like corn, soybeans, groundnuts, etc., have had their genes altered. The scariest part of this situation is that authorities in charge of our food supply, for whatever reason, do not believe these GMOs are harmful. Their conclusion is based on improper investigations and lobbying manufacturers to ignore clear scientific evidence. Can humans do better than the Creator? No!

It is now known that GMOs are not natural. So, what is the problem? Simply stated: they affect our body's metabolism and immune systems. Remember, most of the immune systems reside in the guts. Also, remember the gut is the primary channel through which the food we eat goes into the parts of the body where the nutrients are needed. And finally, our natural foods are specially designed by the Maker to 'fuel and maintain' both human and animal 'machines' called the body. When He designed the human body, He knew the type of 'fuel' suitable for His "machine." He also created ways for His 'machine' to identify and ward off foreign invaders so that the body fights back with allergic/autoimmune etc. responses.

On closer examination, you can guess correctly how the body will react when it encounters GM food—an unnatural food. You cannot fill your car's gas/fuel tank with kerosene oil or water and expect the car to work as designed. Can you? Something terrible will happen! As soon as the brain interprets the type of food eaten as a foreign (unnatural) object, it immediately instructs the immune systems to mount an attack on the alien body.

The consequence? Inflamm on! Thorndike-Barnhart dictionary defines inflammation as "a diseased condition of part of the body, (generally) marked by heat, redness, swelling, and pain."

Yes, it may be possible to drive a car using 'water' instead of fuel or petrol. Who knows! But it must be a different type of car designed to use water as fuel. The Creator created human (animal) bodies, including plants. He designed the type of food that can fuel the systems. Any food can't do. To be able to use GMO foods will require specially designed human bodies that can use GMO foods. Nothing in nature Shows that such a design is possible.

And quoting the Director again, " . . . . recent medical research shows that hidden inflammation, the kind that originates in your gut, is at the root of all chronic illnesses we experience – conditions like heart diseases, obesity, diabetes, dementia, depression, cancer, and even autism." Yet, we ask why the suffering and unhappiness we encounter here on our beautiful earth. Humans ignored the Creator's laws and instead pursued their fancies! The Creator's designs are not to be interfered with without incurring consequences.

GMOs have also found their way into the milk, which almost every family drinks! Because of the quest for more and more milk for "a growing population" or (could it be because of greed?), milk-producing farmers regularly inject into their animals a genetically-altered growth hormone called rBST. This substance boosts their milk production. Before you read further, please google "GAPS ANALYSIS" Report, a Health Protection Branch, Health Canada, and rBST internal Review Team. So, what is wrong with using this growth hormone if it can increase milk production? Everything! Several scientific studies conducted at many world-renowned medical centers like the Harvard Medical School, Kaiser Permanente in Oregon, etc., proved that this milk-producing enhancer causes different types of cancer like breast cancer, prostate cancer, etc. (*Please google this information*). You will not be surprised then to learn that all the countries in the European Union, Canada, Japan, New Zealand, and Australia have banned the injection of this hormone into their cows or any cows producing milk for their citizens. Is the cow producing your family's milk injected with rBST?

In the US grocery stores, all bottles containing milk for sale are marked: "No significant difference has been shown between milk derived from rBST-treated and non- rBST-treated cows." *Check that out*!

While on this subject, we can also look at other things that can affect natural milk or beef quality, making them unnatural – what we feed the cows and pasteurization of the milk.

Remember the command to all "adult" living things, "Go and get your food......" This does not mean any food but the type that suits the body as designed by the Creator."

Imagine what will happen if humans, instead of going for and eating human food (the Creator designed for the human body), decide to go for and eat cows' food or snakes' food or food for pigs? Also, think of the disaster that would occur if humans were to be locked up in comfortable rooms day after day and night after night, fed with the best food available but not allowed to leave the rooms in other words, not free to, "Go and get . . .", i.e., leaving out air, sunshine, exercise, etc. The cows that supply our beef and milk receive that treatment. The All-knowing Creator knew why He decreed that all 'adult' living things must go (move) for their food. Of course, we know that if there were no good reasons, He would have attached the food on the adults as he did with the 'seeds.'

Worst of all, cows that produce commercial milk and beef are always locked up in feeding pens, injected with antibiotics and hormones, and fed and fattened with grains. Cows are grass-eating, not grain-eating animals. Please take note of this. This restriction of movements to these animals and feeding them with food the Creator did not design for their bodies affect the composition and the quality of the beef and milk they produce. It does not require rocket science to come to this conclusion. The consequence of this inhuman treatment of these animals is the inflammation that results when the beef and milk or both are consumed.

Let's turn to Pasteurization. This is a process patented by a man called Louis Pasteur. In this process, raw milk is raised to a specific predetermined temperature and for a certain length of time to kill off harmful bacteria responsible for diseases like typhoid, tuberculosis, etc. This process served and saved the world by eliminating or reducing the incidences of those deadly diseases. Unfortunately, the unintended consequence was also the killing of good bacteria and enzymes, as has been confirmed by scientists.

The Creator designed milk to be a whole food to the recipient and consumed raw (google the composition of raw milk). You would imagine that, with the level of modern sanitation, availability of advanced Health Care and the delivery, and the knowledge of the destructive effect of high temperatures on raw milk, humans, should have figured it out (or have they not?) that it was not the intention of the Creator to have milk raised to high temperatures or doing so was no more necessary because of available technology?

Some critics might still question how the Creator's intention could be figured out since it is impossible to speak to Him to ascertain His intentions. The answer is simple: First, we've seen that pasteurized milk does not digest well with many people due to the destruction/alteration of the composition of the milk. Secondly, if this observation is not satisfactory enough, then we use our usual observation method to provide another answer: Check out what happens with all other living things—in the air, water, and land. In no case is the milk been heated to a high temperature before it is fed to the "young" ones or consumed. Therefore, any milk heated to high temperatures (processed milk) would not be natural but alien. So, the ingenious work of Louis Pasteur should have been used as a temporary fix, not a permanent one.

It has been discussed exhaustively, elsewhere in this book, what happens when the brain senses a foreign body – an attack is ordered and immediately mounted – inflammations! Don't we cry to the Creator, 'Why me,' when some babies or even adults fall sick after a drink of a pasteurized milk? Imagine for a second the type of unnatural food produced if the pasteurized milk is also skimmed?

What is skimmed milk, anyway? It is the leftover after fat is removed from the milk. "And so, what is the problem," you may have begun wondering again? The Creator designed milk (human and animal) to supply, among other nutrients, the all-important vitamins A, D, E, and K. (*Please google this*). Our bodies can't absorb these essential vitamins unless accompanied by fat because they are fat-soluble. Then, you can conclude that the Calcium in the milk will not do its work of strengthening the bones if not absorbed.

You can appreciate the wisdom of the Creator in including the fat in milk. It is clear skim milk is not natural and, therefore, will always be rejected by our bodies, causing inflammation. All attempts humans to re-engineer the perfect work of the Creator will always result in disaster—disease, and unhappiness on this earth!

We know it was mostly the insatiable appetite of humans for more and more that fueled the need for chemicals on our farmlands. The same reason caused the use of insecticide and pesticides on our crops, fruits, and vegetables. Also, include the use of growth hormones on our animals. These unnatural additives are unsuitable for use by God's machines – the animal bodies. Or was it greed, unnecessary waste, or depletion of resources? Could it be that the Creator did not make enough? Let's explore further.

Is it even healthy for one person to consume a whole bottle of milk or one chicken all by himself or eat ten eggs a day, or only hydrate with only milk, excluding water? These unacceptable behaviors are why the quest for more and more started.

Should adult humans consume milk and or milk products? Nothing in nature shows they should. In its natural form, milk is a complete meal specially formulated by the Creator for the 'babies.' For babies, it is a perfect, fast-growth food in its natural state. Even babies reach a stage in their development when they don't care for or need milk. Then they have reached the stage when they can get and enjoy getting their nutrients from regular food. This is a fact.

The Creator allows the production of just enough milk for the baby – human or animal. He made it that a female's milk glands stop milk production when the young one reaches the stage when it does not need milk. If adult humans/animals need to consume milk as a food, the Creator would have so-designed it.

Another question in this line of thinking is, "Should human babies or adults consume animal milk?" Nothing in nature shows it is the intention of the Creator for humans to consume animal milk. Granted, milk, animal, or human, is nutrient-rich. But animal milk is specially formulated for animals and humans for human babies. In no situation have humans fed human milk to animals.

Some might argue that animals are not fed human milk because of scarcity. That is not true. If the unnatural things done to these animals to produce more milk are done to humans, they will produce more milk. A female, human, or animal is designed to make milk only when she has a baby. So, forcing an animal to produce milk when the animal has no 'baby' is unnatural and carries consequences.

Some people might argue that milk does not kill. After all, humans eat animal meat without experiencing any danger. This might sound true. But are they? No! It would be like saying that consuming zero-Cal drinks, including the added small amounts of carbonic acid, don't kill. They can say the same thing about breathing in coal dust like the coal miners, smoking cigarettes, consuming alcohol or hard liquor, allowing small amounts of radiation, etc. No, they do kill! The poisonous effects build up over time, making them slow killers.

As explained above, milk is not for adults – humans or animals. The price humans pay for contravening the Creator's law is the sickness that follows milk consumption.

Other animals and plants take just what they need. A lion, when and only when hungry, will go hunting to satisfy the hunger. He does not kill eight animals, eat one, and put the rest in the refrigerator for later. Of course, in that case, most of the food saved for 'later' will rot and be discarded. So, he will go for another killing spree? It would be unnatural for any lion to operate outside his normal behavior.

Do animals disobey the Creator's laws? No! Only humans do as they please. How many fish does a family need? Just enough to satisfy their hunger and nutritional needs! But think of what goes on here on earth that leads to the depletion of natural resources. Excessiveness results in the desire for more and more necessitating the use of chemicals. To keep these foods longer than required leads to refrigeration, canning, and packaging. These methods are not natural, and hence, there is a growing number of problems associated with consuming them.

In canning, food is generally subjected to high temperatures, destroying most of the enzymes and nutrients that were intentionally put in them by the Creator for different functions in animal bodies. Various preservatives are also added for prolonging shelf life. Finally, different 'natural' food colors and tastes are added to appeal to your sight and test buds, making the product impossible to resist.

Bottling, particularly soft drinks, is another area of concern. A good example is the bottling of orange, mango, guava, pea, etc., juices. Here the juice is squeezed out of the fruit and the fiber discarded. Then preservatives and other foreign ingredients to enhance color and taste are added. After this processing, the juice is no more in the form God created it in.

Fruit juice is not only food but also medicine with its content of vitamins, minerals, enzymes, and fiber. The discarded fiber is designed by the All-Knowing, to among other functions, aid digestion and strengthen the digestive tracks and the colon. Then what are we left with after the processing? Food stripped off its disease-fighting qualities.

Another concern is bottled soda drinks, particularly diet sodas. What is soda? Almost all sodas start their existence with carbonated water (acid) and any of the myriads of food coloring, including preservatives to increase shelf life. The most offending ingredient is the sugars (or more accurately processed sugars), which are in disproportionate quantities. From the Sugar Association, www.sugar.org: regular sugar occurs in nature and, therefore, a healthy part of our diet. As nature's preferred sweetener, occurring in nutrient-dense fruits and vegetables, sugars are the primary sources of the body's fuel for brain power, muscle *Cal Mbano* energy, and every natural process that goes on in every functioning cell.

Generally, an eight-ounce soda contains about eight cubes of processed sugar. Small amounts of carbonic acid may not be a big concern. But consumed over a long period, many health problems like diabetes will start. Processed sugars with their high carbohydrate contents, in any of their different forms, are known to be the leading cause of obesity in children. In many families, particularly in advanced countries, soda is occasionally the main hydrant consumed and water.

Due to the outcry over-processed sugars in this product, manufacturers became creative and decided to respond to customer concerns and increase their declining profits by introducing diet sodas. In this new product, the natural sugars are substituted with artificial sweeteners like saccharin, aspartame, sucralose, etc. Note that these 'no-Cal' sweeteners are lab-created products, not natural. Saccharin, for example, was stumbled upon by a chemist experimenting with coal tar, a carcinogenic material.

As Rodale's Organic Life, www.rodalesorganiclife.com, explained: the problem with 'no- Cal' sweeteners was that when they were consumed, the signal the guts get is that a high-calorie food is on its way. A problem occurs when no high-calorie food arrives— the guts don't utilize the foods efficiently as God designed them. Secondly, the body releases insulin when it receives the signal as if sugar was eaten. You can see why this insulin leads to blood sugar spikes and why such artificial sugars are not suitable for diabetics' use even though it is so marketed.

It is almost unimaginable that a child with its 'milk teeth' not yet shed would risk death from anesthesia during surgery to remove a cavity. "Cavity" in a child of nursing age? Incredible! These are the effects of consuming excessive amounts of highly processed and no-Cal sugars.

God indeed gave humans the brains and their senses to go where no man had gone before. Yes! But only to take care of not to alter His creations, as explained in the above analysis.

The consequences of producing and consuming processed foods are Auto immunities, inflammation, cancer, cavity, etc. – unhappiness!

The Creator made the earth the happy home for all living things by meticulously providing all they would need. Many things have gone wrong due to the unauthorized activities of humans. We are sure it is not from the Almighty God, the Omnipotent, and Omniscient. The Creator God is incapable of making mistakes and cannot forget a thing. It is humans and their new ideas that have brought all the misery to the world.

When God created all living things, He provided for them, all they needed to fulfill the purpose for which the Creator created them. He gave the brains (intelligence), the seat of wisdom, and the senses to humans only. We can confidently conclude that then the Creator established His laws to guide humans. And He coded these laws in the innate human drives for finding and eating the food for nourishing ourselves, keeping us healthy, and aiding us in fulfilling our roles, using the water, breathing the air, and absorbing the sunshine.

Also, we must use the fertile soil provided for planting and sowing. Finally, we are commanded or driven to protect ourselves and what is ours and extend our type through procreation.

To humans only, He added, "You take care of my earth including all in and on it." By studying nature, it will be evident that this was the way He intended the earth to be and function. The Creator never needed help from humans.

All these divine blessings notwithstanding, humans still disobey the Creator's laws, bringing on our otherwise beautiful earth its current miserable conditions.

I will explore a few other activities of humans, not other animals and plants, that have led to the strife and unhappiness here on earth. These are Exploration and Migration, including the forceful occupation of alien lands. From all the evidence abundant in nature, we believe there was the command, "Go and get' your food – interact, seek help and acquire more knowledge from others.' The Creator›s intention is not for any person or groups of persons 'to go and take' what belongs to others, deplete other peoples' resources or occupy lands that belonged to others.

How can we be sure of this? For starters, He distributed his resources equitably among His creatures wherever they are located. Equitable here means that the Creator knows best what and where to locate any resource. Anything he does is good and for the good of humans. The 'Go and get . . . ', and the fact that He did not give one person or group of persons more brains and senses than the rest, but enough to serve their needs, show that He intended to foster communal living, dependence on one another. But every individual, ultimately, will have to make his/her own decisions. But over the centuries, humans have consistently violated those laws. The consequences of which are all the suffering on earth.

Excellent examples of "Go and get . . ." are the outcomes of distant travels. Let's investigate whether traveling violated the Creator's law.

When distant travels to foreign lands are discussed: Henry the Navigator and his marine engineering school; Columbus expeditions; the Magellan team, the Wright Brothers, etc.,

come to mind. These people made their names in inventions, explorations, and distant travels.

Before Columbus reached the Americas, a trade route had been established between the Europeans and the Far East, particularly for spices. West Africans had already started trading with American Indians and Europeans with West Africans, particularly gold and iron wares. So far, everything was fine, just as the Creator intended. They got to their destinations, Exchanged (bartered) goods, exchanged knowledge, and went back to their home of origin or settled among the peoples.

As discussed previously, immigration will always involve compromises and consequences that impact both the indigenes and the aliens. The interaction between these explorers and host natives generally started with exchanging goods or looking for raw materials. Before the indigenes knew what was happening behind the scene, their land was occupied, usually forcibly. What happened in West Africa, in particular, and the rest of Africa, in general, are prime examples.

The Europeans arrived in West Africa as innocent and harmless traders. But as soon as territories opened in the Americas, demanding experienced ironworkers and people who could withstand the harsh environments in the new land, West Africans particularly were herded away in droves to the new territories. In most cases, they were necked in chains and put into slave ships. These unfortunate Africans were subjected to the most inhuman atrocities ever. Just think about this for a moment! The Africans did not invite those Europeans or provoked hostility. They did what was natural to them – welcome them with open arms.

Stories of how sharks trailed slave ships because they feasted on slaves thrown overboard (dead or sometimes alive) during the journey still hunts human conscience. It is also known that when slavery was outlawed by the British, slavers on the high seas used to empty their human cargo into the sea to avoid detection and capture as soon as they sited approaching British slave monitors.

Africans that survived the torturous journeys were promptly put on auction blocks and sold to the highest bidder upon reaching the new lands. Then started the dehumanization of the Africans and the mental realignment program of who they were and the purpose for which they were created.

The story of slavery in America cannot be a chapter in this book or any book. It will always be many books. I will refer you to "Roots" by Alex Halley or "Uncle Tom's Cabin" by Harriet Elisabeth Beecher Stowe, to name a couple for further reading.

The problems resulting from the permanent mixing of Africans and Europeans in America remain today. The Creator made the homes of Africans in Africa, not in America or any other place. After Africa was scrambled for and divided up among the European powers, they began the physical occupation of the lands and the psychological occupation of African minds. All in all, the main aim of the invaders was to secure uninhibited freedom to exploit raw materials to sustain their Industrial Revolution (or civilization).

Does anybody need proof to ascertain that this was not the intention of the Creator? That whatever the Creator put in any environment is for the people of that environment to use equitably. The provision of resources: crude oil, minerals, and precious stones, or even farm products in different places and different quantities was to foster communication and interdependence on one another.

Nothing in nature shows that the Creator intended to foster or encourage greed and exploitation. Trading, like one established between Europeans and the Far East on resources like the Spices, was on a mutual agreement and benefit. In the mighty wild forests, all animals created to be in that forest, big and small, coexist, each species minding its business.

Problems will occur if an animal strays into another's territory. Unfortunately, some humans interpreted His command, "Go and find what you . . . ." just to suit them – to, "Go and usurp what belonged to another." Hence the unhappiness that will ever exist among the peoples of those places.

Ocean-going vessels provided means for transporting goods and for long distant travel. The invention, cars provided pleasure and style in short foreign travel. Trains or airplanes supplied the means for both short and long-distance travel and means for moving goods. Humans have gone to the Moon, and plans are underway to send humans to the planet Mars. Everything appears alright except for the pollution associated with the type of fuel used to power these vehicles and the strife resulting from the exploitation and use of fossil fuels.

A good question would be, "What is the purpose of these technologies and the journeys made possible from their uses?" Are they necessary – the technologies and their uses? Did they make humans more peaceful and happier as the Creator intended? Certainly not! People leave their homes for distant lands for different reasons – mainly to trade, learn/ understand each other's ways, find/discover new things and places, some for pleasure. In some instances, people were forced to journey to and live in other lands, as explained earlier.

How can we be sure the Creator sanctioned these journeys? First, no person is given the knowledge of everything, or all the human needs located in one area. Consequently, interactions among people became necessary. In this process, we acquire and share knowledge from which man's lot is advanced. Other animals such as birds, land, and sea animals do not benefit from their interactions to improve their state in life. Why? They were not given the brains – intelligence—as humans. Therefore, it is certain that the Creator sanctions the journeys as far as acquiring new knowledge through interaction and satisfying different social needs.

He provided humans with the brains – intellect, to achieve these and materials to build the vehicles. However, such journeys were supposed to be temporary and carried out only when needed. How can we be sure this was the intention of the Creator? Take black people, for example (or any ethnic group of your choice). God created them, gave them their distinctive features, provided all they needed in their environment, and placed them in the earth's equatorial regions.

Examine the type of food they eat – one specially designed to suit their systems. If the Almighty Creator put all one needs in his environment, barring forcible circumstances, why would there be the need to move permanently or relocate into another place? Consider the damage man has caused to the ecology of the ecosystems by transferring plants and animals from one to another, where they did not belong? Even when adaptation is possible, systematic problems still occur. We can also consider what happens with other creatures of the earth – those that live on land, in the sea, or air: we know that some sea animals and animals that live in the air migrate during certain times or seasons of the year.

We also know that this move is generally a temporary, not a permanent, fix to find food during times of food shortages. These animals go back to their Creator-assigned homes as soon as the reason for leaving their homes ended.

Elephants, lions, monkeys, etc., are found in their creator-assigned places. They don't move and occupy any location they choose. No! Or else agile and powerful animals like lions, tigers, etc., will be found everywhere on the surface of the earth. No! They know their assigned places and obey the Creator's commands. It is at a high-cost animals can thrive or reproduce in new environments of capture. The Creator established the rule for all living things, including humans, to occupy specific territories that contain all their needs. Did humans, the mighty, and the powerful accept that? No! The result of this behavior or disobedience is the discontentment and suffering we see all over.

When peoples move about to interact as the Creator ordained, there would be an opportunity to be attracted to things they might want to incorporate into their own needs. When Europeans traveled to Africa, they took with them their type of religion, schools, medicines, their diseases, their kind of justice system, and relationship types. You would think that those would help them in their journeys and temporary stay in the alien lands. Did they return to their homes, or did they decide to seize the new lands? You know the answer!

Were these types of transfers to host countries necessary? Were the Africans better off with these new ways of doing things, methods of the Europeans? Every fair-minded person knows the answer is 'no.' The Europeans were supposed to provide some conveniences to support them on their journeys and the temporary stay in the foreign lands. They were not supposed to stay permanently or leave whatever they took with them there in the foreign lands if they cared anything about the laws of the Creator.

Africans had their own culture; different ways of worshiping the Creator as they learned from their ancestors. It defies reason why anybody will believe or teach that there is only one 'correct' way to worship the Creator. As I had already narrated: when I was growing up in my little village, there was order in the society. Also, there was respect, happiness, and above all, reverence to God among my people. These disappeared as soon as Christianity was introduced. Now they learned that they needed not fear to commit any sin or to displease the Creator. God is 'good and kind.' So if you confess your sins to another person, the crimes would be forgiven.

You can imagine the effect of this new creed on the Africans and their culture. Whatever the foreigners took with them from their own cultures was good for them but not for the Africans or other cultures unless what the Africans decided on their own to adopt.

Think of what happens on a typical Sunday (for Christians) or Friday (for Moslems). It is on these two days that Hypocrisy reaches its height. People are segregated according to their religious beliefs and socio-economic status; when children learn who their enemies were, who they could associate with or not.

Compare this to what happened in my village or most African communities then: Friday and Sunday were not the only days of worship but every day. How did those simple-minded villagers worship? Their prayers lasted for a few minutes and were very simple; first thing in the morning and last activity at night. Through their intercessors, they usually acknowledge the existence of and their total dependence on the Mighty Creator. Finally, they declared that they understood that His will must be done – the spirit of 'live and let live'; it's their duty to procreate, rely on, and use all the gifts bestowed by the Creator equitably.

You can see nothing is missing in that simple prayer. For the Europeans, their ways were right for them, and the Africans' ways were suitable for the Africans. This way of praying changed when the new method, preached as "the only acceptable way" to worship the Creator, was forced upon the people: avoiding sinful behaviors at all costs was no longer necessary; the respect and family order, fear of God;

resistance against divorce, selfishness, lying, cheating, murder, became things of the past. Hence, the unhappiness we see everywhere.

The main problem with the transfer was that the Europeans' imposition of themselves and their cultures on the Africans gave the Africans no choice. It has been stated earlier that Africans had their own type of lifestyle – their way of life. It is human to copy what you like, including replacing what one already has with something new or better.

Imagine how peaceful it would have been if Africans were allowed to choose and copy what they liked about the Europeans as these 'visitors' were studying them and copying what they desired of the African's ways. Then, they can modify them as they see fit and rebranding them on their own.

When they arrived in West Africa, they met trained ironworkers and a functioning system of governments. The ironworkers became in high demand immediately after the new world was established. In North America, The May Flower group could not have survived the harsh environment in the new land if the American Indians did not give them lands, showed them how to plant, and deal with the severe weather. The Hawaiians received Columbus and his crew with open arms. Imagine the harmony that would have existed between these two cultures if the explorers had returned the kindness they received from the natives?

Because of the Vatican's issuance of the Papal Bull, the Inter-Caetera in 1493, to Christopher Columbus, what the Indigenous people got in return for their kindness was the massacre of thousands of their people and the seizure of many of their lands. Are you still guessing what prompted Pope John Paul II to ask for forgiveness when he visited Goree island in 1992? (*confirm the events by googling them*).

Europeans brought their type of education and healthcare systems with them. Was education going on in West Africa, for example, when the Europeans arrived? Yes, that is the answer. Young men and boys learned from their fathers the art of agriculture, including how to herd their animals, family life, and husband and a father. A young man stayed by the father's side at the village meetings to learn their justice systems and how to be a leader, including understanding the criteria for receiving village titles and leadership roles.

The girls learned from their mothers how to take care of the family and self, manage resources provided by the father, and become mothers and take care of the children.

At a certain age, she could follow the mother to the women's village meetings where she would listen and learn from old hands. By the time she reached the marriage age, she was ready to face the world.

Be honest with me and forget for a while the "I am better or on top of the world" mentality: What else did they need to learn or be taught; how to visit one of the Stars, journey across the Milky Way Galaxy, relocate to the planet, Mars; produce the most-deadly atomic weapons; discard all peaceful human relationships; how to manipulate other human beings to take advantage of them; how to take care of yourself and ignore the rest of the world? Would any of these bring peace and happiness to our earth, the very intention of the Creator? But these are what European Education teaches.

At home, the boys learned why 'your word' is like your life; how to be a responsible man. After he attended the manhood training (now discouraged by the European Education), he is ready to face the world. Before this time, he would have been in an apprenticeship like learning a trade for his future livelihood. Africans were very industrious! Before Columbus reached America, West Africans were already trading with American Indians, as stated earlier.

In general, girls and boys learned the responsibilities of men and women, particularly their duties to the family and society. Taking care of the environment and the elderly were the most significant responsibilities. No elderly are left on their own. This is what is meant to be human – to take care of those that brought you into this world and took care of you when you could not. The environment was part of their lives; they learned. Men and women were humans but with different roles to play. One would not envy the other but be satisfied with who they were. But with European "me/self, first," we lost what made us human.

Medicine and Healthcare were handled by Herbalists like my father and Native Doctors. Europeans believed that their method was superior to the natives' because theirs were 'scientific.' In some cases, the indigenous techniques were more effective than the European versions. For example, broken bone cases received better treatment and achieved better results with native medicines, even to date. Many critical cases of broken bones are still referred to the native doctors when patients refused amputations (the European method).

In some parts of Nigeria now, some European and native medicines collaborate. There is no question that if the native methods were left alone (not interfered with or discouraged), they would have progressed like any other. Necessary improvements would have followed the natural order. What is encouraging now is that each of the methods is used, whichever is more effective.

Advances in medicine, health, and healthcare delivery have achieved unprecedented milestones. But there is a problem with all the success stories - the use of chemicals as the only way to treat and cure diseases. Then add the effects of the 'get rich overnight' mentality that leads to indiscriminate use or over-prescription of these chemicals.

Chemicals are foreign substances to God's machines – the human and animal bodies. Our immune systems see these alien substances as enemies and would not wait to mount an attack. The result is generally debilitating and sometimes deadly diseases like cancer, diabetes, osteoporosis, etc.

When any disease is treated or cured using a chemical or what we fondly call drugs, it's like killing mosquitoes in a room using chemicals (insecticides) but causing allergies and headaches to the person sleeping in the room. Then you are forced to find some other drugs to treat the headache and allergies. Where do you end the rat race? Nowhere! Because a rat race goes in a circle. While the race is going on in a never-ending loop, inflammation is constantly generated.

I'm sure you have seen somebody who went through Chemotherapy to get rid of or manage cancer. The first thing to notice is the person goes bald – lose all the hair. At this point, all the immune systems and defense mechanisms provided by the Creator, are disabled predisposing the patient to be affected by all types of diseases. The only choice would be to administer other chemicals to counteract the side effects. Don't forget that the drugs counteracting the side effects of the first drugs have their side effects. You get where I'm going with this.

A friend of mine was worried about his bald head. He believed that would affect his looks. After a few weeks, the daughter announced that she found a cure for the bald. My friend was delighted with what he described as 'God-sent' good news. But before he could take the drug, he searched for its side effects. When he learned that one of the side effects of the drug was 'developing breasts,' he decided to accept the bald condition.

Imagine the amount of radiation the body is subjected to during x-rays – any x-ray. The radiation in large doses is deadly. There is a department of Anesthesiology in most modern hospitals. Here heavy drugs like the Propofol are generally used during surgeries like the brain, by-pass, amputation, C-section, etc., to lessen or eliminate the unbearable pains associated with surgery.

After discharge from the hospital, a load of drugs will escort the patient home. Remember the 'rat race.' Can this heavy load of drugs be avoided? Impossible! It is an integral part of modern medicine! Try to take a peep at the Medicine Cabinets in many homes. You might think you were looking at a section of a drug store. You will not know what the drugs are for – to treat ailments or for counteracting the side effects of the primary drugs.

The proper question is, "How can surgery be performed considering the pains associated with the procedure?" Give the patient a few shots of whiskey and strap the person down with heavy ropes and start cutting and hope for the best? Can you imagine applying this crude method during heart surgery, brain surgery, C-section, etc.! The resultant wailing will wake up a whole village. This is where the miracle of modern medicine comes in. The anesthetic should be described as a blessing to patients requiring these operations.

Tooth and bone problems are prevalent among younger and older people. Luckily, there are the Dentists to handle all our tooth problems or Orthopedists to deal with the diseases of the bones and joints, etc.—too many areas of Science and Technology dealing with the health and wellbeing of living things.

Going by all these, it is easy to conclude that humanity was not taken for a ride by civilization but rather profited from it. But not so fast! This conclusion ignores the outcomes of due diligence on "cause and effect" in modern science and technology. At least in this book, it is not feasible to examine all the branches of Science and Technology, leaving me with only one choice – cite one example to make my point.

A simple toothache will send a person to a Dentist for diagnosis. The verdict could be 'Cavity.' At the end of the set of Milk Teeth, every human must acquire the permanent set of adult teeth. The sets are perfect as created by the Creator of all things. But how and why cavity? Scientists have shown that too much sugar (mainly processed sugar) will always lead to tooth decay, which will result in a cavity. Then science and technology are summoned to provide a cure.

Humans must be grateful to Science and Technology that are obligated to 'run' to rescue them from the consequences of disobeying the Creator's laws. As prescribed by God himself, the only food we must eat (put in God's machine) is the very one He created Himself. No! Humans must find a way not to obey. And by eating the forbidden foods, inflammation is made in our bodies, which will eventually lead to diseases (cavity), sickness, and death.

Not only the wrong foods we eat that cause sickness and death but also the chemicals we use as drugs. It is forgotten that the Creator's machine – all living things' body, is designed to accept only natural elements. Drugs are chemicals and, therefore, not natural. There is nothing new about using medicines to manage or cure diseases. After all, doing that started with the Creator Himself. And Medicine men and women learned from Him.

The Creator put His in the food we eat. He does not need to send His angels to our earth to administer medicines.

Must chemicals be used to cure diseases? Not really! Did the Creator provide natural cures for illnesses? Of course, yes. The Creator is All-Knowing and can't forget a thing. Check out the medicinal contents of plants and vegetables; fruits, seeds, nuts, and oils; meats including fish; dairy products and crops. Many of our current medicines have their bases in plants and animals. It is the way the drugs are extracted and manipulated that will turn a simple harmless natural herb into a chemical substance.

Then, there should be no doubt that all diseases can be cured naturally without the use of chemicals. How can the accuracy of this statement be verified? From all that was discussed previously, we know that the Creator does not want chemicals used in any living thing – His creations. Being the Ever-Knowing Creator—He knows everything, past, present, and the future, and He does not take chances. He put all the natural remedies for diseases in the fruits, seeds, vegetables, herbs, etc.; he created – the food we eat. He knew diseases and sicknesses would be a part of living or, at least, for a time.

If you have not experienced the disastrous side effects of drugs personally, you may have seen a person debilitated by them. I'm not referring to an isolated case but common occurrences. Once you start taking drugs, you run the risk of entering the 'Rat Race.' Indeed, He does not want chemicals in animal and human bodies.

The right question is, why do humans prefer unnatural cures or chemicals to natural remedies (no, manage diseases)? Could it be ignorance – not knowing what the Creator wants? Or is it because compounds yield more money? I prefer to guess both. Because humans are ignorant of the time element built into our lives – everything needs time to come to fruition. So, we become impatient with the slow healing process of natural cures. Chemicals work faster than natural remedies. But our bodies see and reject them as unnatural elements, causing inflammations, diseases, and death.

A time is coming when humans will be tired of the 'Rat Race.' Then, the Big Pharma and Drug Companies will pay more attention to the Creator's herbs to discover powerful medicines that will cure all diseases. Some may have been discovered already, but greed or the quest for more money kept them hidden.

Then knowing what these chemicals are – unnatural substances, whose use is critical, at least for the time being in curing or managing diseases, humanity is left with only one choice: Use the chemicals sparingly and, most importantly, as a last resort. In time people will discover natural cures.

This chapter started with the question, "Are we taken for a ride by Civilization?". It laid bare the pros and cons of modernity. It is apparent humans have advanced in all spheres of life. You would think that all the achievements in Science and Technology would have ushered in peace and happiness throughout the world. But we are far from reaching that milestone.

From the analysis described above, we know the Creator intends to have a Paradise Earth. Unfortunately, the earth is anything but a paradise. What is the matter?

First, Science and Technology failed to deliver the promised land, instead left us with insurmountable problems – the by-products of civilization. These are polluted water, air, and land; we are stuck with unnatural substances (chemicals) to cure or manage diseases caused by those chemicals.

The food we eat did not escape pollution, considering the use of growth hormones and other chemicals on our animals, making milk, meat, dairy products, poultry, fishes, unacceptable to our bodies. Instead of aiding the metabolic functions of the body, our food causes inflammations, and diseases. Also, consider the following materials that find their way into our food.

Such products are fingernail polish, chapstick, shoe polish, hair spray, air fresheners, metallic and plastic coating substances on our cooking pots, plates, and utensils. This list includes food colors and preservatives, kitchen, and bathroom soaps; plastics; processed and refrigerated food items, bottled, canned, and diet drinks; artificial sweeteners or sugar substitutes. These materials get into our bodies and cause inflammations and diseases.

Artificial lights have, in many cases, rendered natural nights unnecessary, causing the depletion of the natural vitamin D; nuclear and spent nuclear materials and weapons of different types, capacities, and strengths are stuck with us. Just imagine the number, classes, and quantity of prescription drugs, including feminine hygiene products, flushed down the toilets daily. These drugs will eventually find their way into our drinking waters. Hardly can water purification systems get rid of them.

Should we be surprised to see some young boys developing breasts or a high incidence of obesity, early onset of puberty, etc.?

Soon we may run out of space for our trash – chemicals, non-degradable materials, and spent nuclear wastes. Luckily, recycling has begun but sure cannot keep pace with the amount of trash produced. Imagine, some people suggest that our ever-increasing garbage should be dumped in space, just above the earth's atmosphere. Would you please think for a minute about the disaster such an action will create?

There has been stiff resistance against moving nuclear wastes from one state to another. Even some people would not allow the materials to be transported through their states. Because of this difficulty, there was a bizarre unconfirmed rumor that some people suggested that the toxic materials should be sent to some third world countries where, with few incentives, the rulers could be convinced to accept them. I don't see how such a transaction will see the light of the day.

It is necessary to observe here the difference between chemicals and natural materials. Natural materials are as God created them, like the unprocessed food we eat, plants and animals produced organically, unpolluted water, the air, etc. These are earth materials, and the trash does not need to be dumped elsewhere but where they were produced. They are biodegradable.

But chemical materials are produced in the laboratory by chemically combining two or more elements. A chemical can also be produced by altering the natural composition of a material, making it unnatural. Some examples are plastics, drugs, GMO foods, etc.

Many people have turned to Organic farming in the hopes of producing and consuming foods untouched by chemicals. This type of farming is an exercise in futility. Where can an unpolluted or uncompromised piece of land be found to raise organic foods? Hardly!

A piece of land can be compromised from different sources: Direct application of chemicals – insecticides, different types of spray, etc. Our garbage dumps contain different types of discarded materials containing chemicals that, in time, sip into the underground water systems.

Regularly you must water what you planted in your 'organic' farm. Where will water untouched by chemicals be found? Even rainwater must pass through a polluted atmosphere.

It is Civilization that brought about family insecurities and breakdowns; selfishness; 'I'm better, and on top of the world, the rest are beneath me.' The Internet has made it possible and easier for false prophets and preachers to reach all peoples of the world.

Thanks to Civilization, the concept of and what affords happiness have changed. Happiness is a condition or state of mind. It is internal and can't be bought with money, contrary to the beliefs of many people. We know it is not so.

Before civilization, families used to work in their farms and gardens, producing all they needed. They go home, cook from the products of their farms. There were no chemical products to compromise the type of lands on which they planted.

There was no need for money since they produced all they needed. And the concept that it is money that brings happiness was unknown. Then, they sit around, tell stories, and enjoy the company of one another. The father was there with the wife, the parents and grandparents, the children, and grandchildren — no worries about anything – happiness.

Take an informal survey now on personal happiness. Ask simple questions like, "What is the one thing that will bring happiness to your life." You'll be jolted by the answers you get: a bank account of about twenty million dollars; own two or three mansions in our Capital City; own a private jet plane; buy my third car; get married to a millionaire; etc. Answers like these reveal that many people set their minds on fancy, on things they think will bring happiness, but in real life, things that will bring unhappiness.

This mindset has two outcomes: One could set a goal for themself to achieve one or more of those goals. And throughout the life span, work hard to achieve them, and never succeeds. In the end, discontentment and unhappiness will result. Some people will achieve their goal(s). But happiness remains elusive, considering what usually happens after those things are fulfilled — infidelity, expensive court battles, divorce, loneliness.

All one needs to achieve happiness is already given by the Creator. And they are free. The problem is that the mind is set on the wrong things (stuff), as described above, things that are not part of the Creator's arrangement.

Then how do you handle Civilization and its by-products – the technology, new ways of thinking, and doing things.? No living thing can escape modernity and its effects. To try to avoid them is like jumping into a lake stark necked and try to find a way; the body will not get wet. Therefore, while interacting with civilization, understand what you are dealing with, and be aware of the consequences.

If you like to consume sugar, particularly processed sugar, understand that going overboard with eating it will cause overweight, diabetes, cavity, etc. Then, excess sugar is not suitable for your health.

Chapter 20

PANDEMIC DISEASES

There is yet another source of unhappiness here on earth — the scourge of pandemic diseases, an apocalyptic type of crisis. The flu season has turned into a seasonal ritual.

Already, capitalists have turned it into a celebrated event. Every year we are bombarded with advertisements and incentives (like gift cards) urging us to get our flu shots.

Sometimes the shot may not treat a strain of the virus, and you may have to get another shot.

You may wonder why people simply expect the flu season without asking the critical question, 'Why?'. Why must we get one or more flu shots every year, and why can't humans eradicate the flu? I decided to google US – CDC (http://cdc.gov/onehealth/basics/zoonotic- diseases.html) for the answers. I was surprised or almost shocked by a straightforward illustration that I found relating the spread of diseases between animals and humans (shown below).

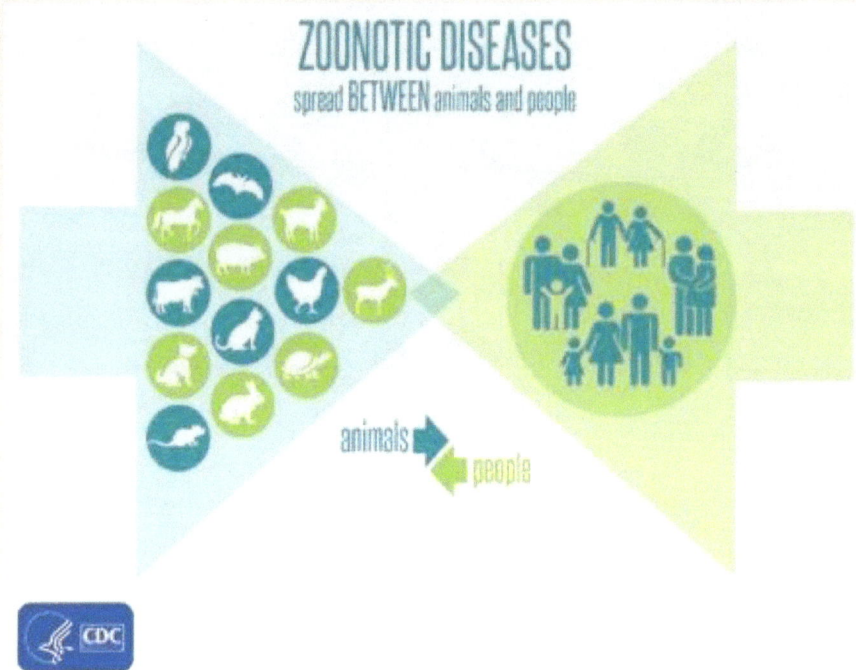

Fig. 9 Pandemics and the spread of zoonotic diseases.

It is a remarkable picture to see! It appears nature plays a role for disease to be transmitted from commonly known animals/pets to humans. Flu viruses are known to spread and cause infections in domestic animals, including wild birds, pigs, horses, and dogs, with sporadic outbreaks in seals, whales, ferrets, and cats.

Vector-borne pathogens are spread to people and animals primarily through the bite of an infected mosquito, tick, or flea. Canine influenza or dog flu is a respiratory disease caused by the type A influenza virus. There has been no known case where the disease passes directly from dogs or any other animal to humans.

For example, bat flu must go through different mutations before it is able to infect humans. This disease is common in Central America. Cat influenza, including avian, infects cats that pass on to one another as humans pass the virus from person to person. Cats passing this flu directly to humans is uncommon.

Note here that due to the mutation abilities of these viruses, they can in the process acquire the ability to pass onto or make a 'zoonotic leap' to humans. That is the only way that animal diseases can infect humans.

Included in what has been described as the usual ways germs can pass from our domesticated animals to humans are as follows: sit on a chair vacated by an infected animal; uncontrolled handling of an infected animal; casual touching of infected animals in zoos or anywhere; eating an infected animal and breathing in the dust in a place where an infected animal was kept; bites from mosquitoes, ticks or fleas that feed on infected animals.

Just imagine for a second what happens when a person kisses a dog, a horse, etc., in the mouth or sucks the animal's tongue as a show of passionate love. Also, consider the type of diseases the animal could be carrying even if the animal appears very healthy? The diseases can easily be transmitted to humans. When a person kisses an animal, the same person will inevitably kiss or come in close contact with other human beings. This scenario is a typical example of how diseases capable of mutation pass from animals to humans.

It is now understood that animal cases of flu (some of which are mentioned above) will always present mild sicknesses. It is only due to their abilities to mutate that can make them dangerous or deadly to humans.

Some of the deadliest diseases are vector-borne diseases. These are viruses and bacteria that are spread by vectors like mosquitoes, ticks, fleas, etc. A virus is a disease-producing agent which depends on the living tissues of the host to grow, multiply and infect. Let's examine some of the viruses:

In 1980 HIV – AIDS, an immune-deficiency diseases visited the earth with a vengeance, claiming more than 35 million lives since 1981. In 2014 36.9, million people tested HIV positive. According to the CDC, scientists believe that the chimpanzee version of the immunodeficiency virus (called simian immunodeficiency virus or SIV) most likely was transmitted to humans and *mutated* into HIV. The disease attacks and destroys the body's defense mechanism — the white blood cells. Human transmission is through unprotected sexual intercourse, shared use of injectors, infected blood transfusions from mother to child during pregnancy, childbirth, and breastfeeding.

Between 2002 – 2003 there was an outbreak of respiratory disease, the SARS. It started in Hong Kong and quickly spread to 37 countries within a matter of weeks. It took 774 lives. This respiratory disease appeared in different strains: first, between

2002 – 2003, then between 2012 and 2013, and currently as covid-19. That strain had a 9.5% chance of death. It is of zoonotic origin (see illustration above).

In 2009, there was another outbreak called The Swine Flu (which contained bird, swine, and human flu viruses). It took 284,500 lives. In the 1957 – 1958 outbreak, it killed 2 million people. It showed up with a fever, chills, body aches, cough, sore throat, runny or stuffy nose, and watery red eyes.

Between 2013 and 2016, the Ebola Virus surfaced in West Africa, claiming 11,000 lives. It was previously detected in 1976 in forest bats.

Once again, humanity is faced with another pandemic virus attack – the covid-19. At the end, thousands of people would have lost their lives to it.

The current epidemic (or at the time of this writing) is the coronavirus or the covid- 19 virus. It started in the wet market in Wuhan, China, as far back as the December of 2019. The symptoms are high fever, cough, pneumonia, and respiratory difficulties. As of writing, it has so far claimed 28,000+ lives with 600,000+ cases. Scientists believe it started in a non-human host and jumped to humans. They have pinpointed that bats, and pangolins (armadillo-like mammals) were the non-human hosts. Pangolins are also found in Africa. The Creator assigned natural homes/ habitats for each of the animals, just like the humans.

These creatures live in perfect harmony with the other creatures of the same habitat. When these animals are forcefully removed or displaced, a natural chain reaction occurs.

Consider the pangolin, the animal whose genetic virus sequence is 99% compared to the coronavirus. Scientists reasonably believe the pangolin carries the virus that made a zoonotic leap to humans, causing the current 2019 Corona pandemic.

For years endangered pangolins have been reported as the most illegally trafficked/ exported animals in the world. Humans have used every part of the unusual, scaly mammal for traditional/spiritual medicine, food, and zoos. For their transportation to other locations (unnatural), these animals are put in boxes/cages and exposed and frightened. Under this condition, the pangolins ooze out a peculiar odor that attracts/ensures bites from other animals, including diseased bats. Coronaviruses are known to circulate in mammals and birds, and scientists have already suggested that nCoV-2019 originally came from bats. Global markets selling live animals are ideal places for such diseases to spread and mutate.

Going by the levels of sophistication in science and technology, this type of calamity will belong to ancient history. But why not? Yes, "Why not"? First, let's consider the number of great minds in science and technology all over the world. Then make a mental count of the number of highly advanced countries on this earth. Among them is America, which has succeeded in sending people to the moon and now preparing to visit Mars.

Isn't it unconscionable that common mosquitoes, teaks, and fleas run amuck in the face of these advances and cause the deaths of thousands of people in any given outbreak? Certainly, the earth has the wherewithal to eradicate these human scourges once and for all. But, because 'we' don't know who the Creator really is, neither do we embrace His expectations of us – to love our neighbors as ourselves and take care of His earth, we chose money instead and allowed human sufferings to thrive.

Some behaviors at the height of covid-19 were reprehensible, like hoarding, price-gouging, or people trying to spread or extend the life of the disease. These were happening while the brave and patriotic first responders were putting their lives on the line.

What lessons can be learned from the pandemic diseases' saga? Many of these diseases and deaths caused by those animals are avoidable.

a) The diseases are transmitted through unnatural relationships between humans and their 'pets.' I have described some of these relationships and given reasons why the Creator requires humans to take care of animals in detail. So, it is not the love shown to animals that is the problem but relating to them as human beings, displacing them from/ destroying their habitats, creating unnatural relationships between the species (like taking wild/exotic animals for pets). Animals are not humans and can easily carry diseases because of their mode of living. So, the distance created between humans and animals by the all-knowing God, must be respected and maintained to avoid these incidences of disease transmission.

b) Infected mosquitoes and bats are the primary avenues through which these diseases enter the human body. Bats bite both animals and humans, passing diseases in the process. Mosquitoes feed on both human and animal blood. This is how they transmit germs from person to person or from animals to humans. They can feed on our blood if we let them, i.e., the Creator has given us the wherewithal to take measures that prevent this from happening.

c) I can say that the spread of these viruses is not of the Creator but of humans. Pangolins, bats, mosquitoes, etc., are creations of God. We don't know what the Creator knows. But we know that all life that He created is good. Also, he gave us the brains, the senses, the consciences and the free will to aid us in the choices we make. We already know how to: vaccinate domestic pets, not take wild animals for pets, and protect animals' habitats in the wild. Our flagrant disobedience to God's laws and even human laws has led to many health crises.

d) Finally, isn't it embarrassing to know that our modern world that possesses the most sophisticated technologies, with great countries, including America that has sent men to the moon and now gearing up for a visit to Mars, will stand by while little animals like the mosquitoes, bats, etc., claim so many lives in every outbreak! It is unconscionable. It is immoral.

Yes, money or the desire for more money is fueling the negligence and ungodly acts.

Of course, to eliminate the menace, you must consider the bottom line of the greedy wildlife trade businesspeople, pharmaceutical companies, and even the medical professionals, individuals that hold the solutions in their hands. The question these people failed to answer is, "Did they forget the expectations of the Creator of all of us and the consequences of disobeying His laws"? At the end of it all, we will go as we came to this world, alone and with nothing"?

Chapter 21

TERRORISM

Please note here that the conflicts described in this chapter may have been ultimately settled, partially settled, or still ongoing. Then why the story? Then why the writing? The cessation of an Injustice does not end the effects. The effects continue to live. Time Out is exploring the causes of unhappiness on our otherwise beautiful earth and proposes solutions.

Another source of unhappiness on earth is terrorism. It is inevitably one of the topics whenever social unrest is discussed. This is because our world has become bedeviled by terrorism as never seen before.

What is terrorism, and why does it exist? Is it an act of God or humans'? In other words, is terrorism due to nature or nurture? As these questions are explored and the answers proposed, an open mind will be required to complete comprehension of the causes and consequences of Terrorism. Please note that I am not a religious historian or a religious teacher, and the writing is not scholarly. I am only interested in the essence of the events I'm describing.

Generally, terrorism is an act of deadly violence meant to attract attention and wild publicity by inflicting maximum destruction on properties or unleashing ghastly deaths and injuries on as many people as possible. The perpetrators of these ungodly acts claim they were revenging the past evils they suffered. They view themselves as indicators for some social injustices, generally imposed by a mightier power, that have no other avenue to a solution but the use of brute force.

There are also splinter groups, self-appointed avengers. In all, recipients of these brutal acts are generally innocent people who have nothing to do with the causes of the 'terrorists' claim to be fighting. How can any sane person inflict pain, injury, or even death, on men, women, and children they never met before? Therefore, this topic is essential and must be discussed whenever the unhappy state of this world is considered.

There is a thin line between terrorists and freedom fighters. The bar is very thin because, in many instances, their tactics and mode of operation are very close. Both see themselves as underdogs of oppression by some overwhelming powers. They concluded that violence was the only alternative to resist oppression.

The violence is not like what is experienced on battlefields, where two armies face each other. No! These fighters are not soldiers in the real sense of the word. They are slippery, hard to detect, and sometimes invisible. They create incidents that simulate the experience of shock, fear, devastation, etc., on unsuspecting, innocent people, i.e., they act out war experiences as groups or a 'lone-wolves.' Examples abound in recent history.

The armed struggle in the Niger Delta of Nigeria

Since the so-called Nigerian Independence from British Colonial Rule, the exploitation of crude oil was placed on the laps of the greedy and heartless multinational corporations, with the notorious Chevron, Exxon, and the most notorious of them all, the Royal Dutch Shell, leading the pack. The Multinationals and the ruling members of the country, Nigeria, enriched themselves from the proceeds of the oil sales. What did the Indigenes, the owners of the lands from which the oil wealth was pumped out, get? Environmental disaster, broken promises, and misery.

These atrocities attracted some verbal resentments from some people of conscience, among which was the group led by Saro-Wiwa, an indigene, poet-turned-activist. No repressive government would accept such a blatant intrusion! In 1995 Saro-Wiwa was murdered under a trumped-up charge to silence the opposition. The result of this heartlessness was immediate – The Movement for the Emancipation of the Niger Delta (MEND) was born in 2005, leading to a change of posture, from non-violent activism to violent resistance.

The MEND's case was simply: that their people who own the lands from where the crude oil wealth came, had a right to participate in the control and sharing of the oil wealth; that it was unfair and inhuman for their people to be the most disadvantaged group in Nigeria; that Nigerian Government should provide reparation for pollution caused by the oil Industries. To be taken seriously, they killed, kidnapped, they destroyed oil processing equipment.

In the face of the support Nigeria enjoyed from powerful countries like America, Britain, etc., that depended on Nigeria's 'sweet' oil, what could the freedom fighters achieve? The people from that area view the Creator's blessing as a curse because the gift of crude oil brought nothing but misery to them.

Ref.: Beloveth O. Nwankwo The Politics of Conflict over Oil in the Niger Delta Region of Nigeria American Journal of Education Research 2015 https://en.wikipedia. org/wiki/ conflict in the Niger Delta.

You may be a Christian, attend church services every Sunday and sometimes during the week. You may be one of those that receive Communion as often as possible. Perhaps you are a Muslim, faithful in attending the Friday Prayers and making your prescribed prayers five times a day. Or rather, you may be a Jew, a Hindu, a Bahai, or a member of one of the Eastern Religions. Alternatively, you may be a simple everyday guy who believes in "Live and let live." Above all, you believe in One Creator, who provided all we 'own' here on earth. How do you see what is happening in the oil-rich Delta region of Nigeria? How do you see the "settlement"? Fair?

Should any indigene be killed or punished for asking for a fair share of God's blessings to his people? Should he be labeled a terrorist for fighting for what belonged to them? Who should be labeled a terrorist, the owners of the land, or the few Nigerian federal government officials who colluded with multinational oil companies? How will the brutal actions of these few shape the thinking and behaviors of current and future generations of Niger Delta people? Such inequities will ever remain a source of unhappiness for those who live or lived in those areas or even remember what happened there.

What do this callousness and inhumanity do to our humanness when we hear of these situations; we shrug our shoulders and then say/do nothing? After all, those children facing starvation in a land of plenty are not ours. Why should we concern ourselves with the problems of 'those' people? Have you ever looked into the sunken eyes of a starving child whose only sin was being born in the oil-rich Delta State?

We all have an obligation here not to rest on our oars until the atrocities end and atonement rendered to the people and their environment. Any settlement based on bribing some few officials who, in turn, will mount a pressure campaign on the agitators will not do. Someday, there will be some tangible form of peace, some change of heart by the Government Officials and their accomplices. Indeed, it is already happening. But how will the effects of the evils committed against these people be forgotten – the murdered children, mothers and fathers, aunts and uncles whose hopes and aspirations were lost? How will future generations forget!

Did these usurpers not sow the seeds of future discontentment and unhappiness? But what we accumulate on this earth will mean nothing in the end. They will be forgotten! We need to go back to nature to understand who the Creator really is and understand our duties here on earth – the purpose for which we were put here.

South Africa

A similar example comes from South Africa. The freedom fighter was Nelson Mandela (1918 – 2013), the leader of a group whose defiance led to the dismantling of the country's racist and brutal apartheid system instituted by the British settlers. The owners of the land were the Blacks of South Africa. The Creator's blessing was not Crude oil but precious stones, diamonds. The South Africans were not only denied the wealth God gave to them, but rather, many were killed for it.

They were herded like cows and goats and relocated away from the wealth, where, if lucky, one could subsist on whatever was available.

Mandela, with his group, could not stomach the injustice. He entered the arena of the Civil Rights Movement when he organized the first non-violent, country-wide protest. In 1961, Mandela became the commander of the newly formed military wing of ANC. This group later morphed into the armed wing of the party. He was later accused of, among others, sabotage and conspiracy to overthrow the South African government. Supported by the US President, Ronald Reagan, and U.K. Prime

Minister Margaret Thatcher, ANC was declared a terrorist organization (sympathetic to Communism) that must be stamped out. Mandela escaped death but paid for his "sins" with twenty-six years in jail. Other ANC leaders paid with their lives.

On their own, Mandela and his group could have achieved nothing. But the world opinion, direct help from many countries, chief among them – the US, Nigeria, Kenya,

Libya, etc., was behind them. Finally, apartheid was dismantled, South Africans reclaimed their land, and Mandela became the first black President.

(Ref.: abcnews.go.com/international/nelson-mandela-dead-south-africa-president-dies/ story?id=8787025)

It is interesting to note how the British bull-dozed their way into South Africa, imposing their rule, mining the precious metal, and herding away the indigenes of the country. How can anybody explain this criminal behavior? They saw a desert land populated by wild animals. Even they counted the owners of the land as a part of the wild animal population.

That was enough reason for them to take over the country and instituted the notorious apartheid system. The British settlers took their religion, particularly the Christian faith, to the Africans. They taught them about the Merciful Lord who 'looked like them.' Being relatives of the Lord, they were destined to go to Heaven.

On the other hand, the Blacks, the landowners, were told that unless they worshiped and protected the masters, they would go to Hell. The natives did not believe them but saw conning usu ers and resisted them. How can bows and arrows of Zulu warriors, among others, challenge British firepower, trained and disciplined soldiers? It was the world's opinion, along with outside help, forced the British to relinquish power to the rightful owners.

Think for a moment about how many South Africans were killed? Many were subjected to all sorts of brutal treatments, including torture, imprisonment, and even death. Imagine the number of South Africans that were raped and enslaved! What was their sin? They were asking the impostors to leave their country and end the usurpation of their natural resources.

These Agitators were also labeled 'terrorists.' Again, the question was: 'who was the terrorist?' The people that matched into other people's lands and participated (even if indirectly) in horrendous acts of taking over other people's lands and natural resources. Or the landowners who were trying to protect their people, lands and retrieve their belongings from uninvited intruders? South Africans were characterized as those who hated "Our way of life." No! They were resisting an invasion of their way of life.

If you are still reading, you know that the Creator forbade any form of usurpation. How? Because He assigned each group of people a place on this earth, complete with most of their needs. Yes, South Africans have recovered their lands and Independence with the help of well-wishers. Surprisingly, these well-wishers included some of the British people who once invaded their lands. How will they wipe off the indelible marks of injustice done to them? The forceful takeover of South Africa by the British demonstrated a lack of understanding of who the Creator is and His laws. Or did they ignore His statutes? They were not taught about the power of God, all His provisions to all living things. They did not understand that without the Creator, they could not even exist. If they did, they would obey His law, "Live and let live," and know that the Creator commanded human beings to take care of each other and the whole earth.

I think the South Africans were wise in establishing the Truth and Reconciliation Program, realizing that revenge and retaliation have no place in South Africa, and they would stand to lose what they gained. The God-fearing white South Africans who listened to the 'small voice inside' cautioning them that Apartheid was inhuman and that usurping another person's things was not of God should be praised for their courage. All South Africans, both black and white, love the land and can live together because the Creator made South Africa spacious enough to accommodate both. The greatest gift to that region was Mandela, who demonstrated by example, the overwhelming healing power of forgiveness.

Non-Violence

Another brand of 'Freedom Fighters' occurred in India under Mahatma Gandhi (1869 – 1948), the leader of the Indian Independence Movement against British Rule and in the USA under Martin Luther King Jr., a model for the Civil Rights

Movements in America. The method and tactics of both were the same. The latter learned from the former. Therefore, they are considered together in this book.

The name of their movement, "Non-violence," says it all. In 1930, Mr. Gandhi led his countrymen and women to match against British-imposed Salt Tax, and in 1942, called on Britain to quit India. Dr. King was fighting against racial discrimination and demanded civil rights legislation to protect the rights of African Americans and other minority groups. In one of his matches in Washington DC, King delivered his famous speech, "I have a Dream."

These people did not only refuse to pick up arms to fight their 'oppressors', but they also resisted fighting back when beaten, water-hosed, or even when sent to jail. But they were resolute in their determination to showcase their grievances to the whole world through disrupting businesses, matches, and legal interventions.

Finally, Mahatma Gandhi and his group traveled to Britain to speak to British people directly. Similarly, Martin Luther King went to Washington and met directly with lawmakers. At last, the former British Indian Empire was granted Independence following the division of the Empire into Hindu-majority India and the Muslim majority of Pakistan.

FOR THE FIRST TIME, Martin L. King and American minorities had a series of laws, known as "Civil Right Laws," that placed all Americans toward an 'equal' path. There was even a time when Martin Luther King was called a 'Terrorist.' Both Dr. King and Gandhi and many of their supporters also sacrificed their lives for their causes. Mahatma Gandhi was shot, and so was Martin Luther King Jr.

Ref.:http://en,wikipedia.org/wiki/mahatma Gandi; www.youthfots/champions/ martin- Luther-king-jr.html.

I will comment on the Civil Rights Movements in India and the USA together since both are, as stated earlier, "Non-violent." It is a fact that India and Pakistan were never parts of Britain as Britain was not part of any other country. Then Britain extending its rule to those places is absurd and illegal. The dehumanizing effects of colonization and the repressive regime of the colonizers are well documented. The never-ending mistrust, suspicion, and hatred instituted by the British before they left Nigeria are prime examples of what happens at the end of colonization. This is what is happening between India and Pakistan to date?

Current and future generations will never forget the evils generated by the temporary occupation of the British Indian Empire and the subjugation of the citizens. Today we pretend ignorance that unrestricted British actions in those countries are the root causes of many despicable reactions from those people.

The case of black Americans was different. Here groups of people were captured from their homelands, not due to acts of war, but by the force of greed. They were shipped to America and European countries under the most inhuman conditions. As if those atrocities were not enough, they were subjected to higher inhumane conditions in America – separated from their families and loved ones. They were displayed on auction blocks and sold to the highest bidder as done to goats and cows. The men were reduced to 'a fraction of a man' through methods like being forced to watch while their daughters, wives, or mothers were rapped.

Slave men and women had no rights to their offspring. Many of the children were sold, turned into foot warmers, and sometimes even crocodile baits. Men, women, and children were turned into beasts of burden and bred like cattle as if they were created to serve the master. These slaves were forced to toil from sunup to sundown in the master's plantations.

Christians, Muslims, God-fearing peoples, 'every-day Joes', would you withstand the types of treatments to which these fellow human beings were subjugated? Have you ever thought about the kind of men and women that inflicted these types of wickedness on fellow human beings?

Were those who inflicted this cruelty not terrorists (at least to the recipients of those inhuman acts)? How would any of the Africans who lost a son, daughter, wife, or husband see those human traders? Even to date, there is a community in Igboland still being accused of participation in slave Trade (notwithstanding all arguments to the contrary) that led to the loss of many of their people.

Ironically, some native-born Africans here in America suffer resentment from a few Black Americans. Some of these Africans, like the writer of this book, are married to Black American women. This show of kinship put little or no dents on the resentments. The competitive environments of workplaces and schools do not improve the relationship.

Just imagine it: back home in Africa, many African families are still grieving over family members lost to the slave trade. However, in America, those same Africans could be accused of participating in the slave trade and suffer resentments as a consequence.

Is it possible that Africans did not participate in the slave trade? That would not be easy to sell. Suppose the current arrivals of Africans were subjected to what black Americans have gone through here. Wouldn't they be equally angry? Certainly yes! Still, the fact remains that the ancestors of every African in Africa did not and could not have sold every person of African descent in America into slavery.

What we need here is understanding, particularly from current Africans here. We've seen what Black Americans go through here daily. New African arrivals should make every effort (bend backward) to bring the Black Americans to close, regard, and treat them as brothers and sisters, which they are. Marrying black women or men is a good start. But we can and should do more than that. Will it be an easy task? I think it can be. We must try and disregard the resentments from some Black Americans.

Up to date, many Black Americans still suffer some of the same types of injustices described above. Yes, they are expected to be pleased people – people whose ancestors were brought here in chains! Yes, it was sold to them that they were brought here, 'a land of freedom' where "all men" were created and treated equally." They were expected to be thankful and happy to have received every type of 'kindness,' including removal from Africa's 'savage' forests and made Christians. Just think about this for a second! 'Be thankful' to those who snatched you away from your home, loved ones and forced their religion and way of life on you! Isn't this a strange world?

Some argue that many White Americans were so generous to the blacks, they even voted for a black man, Barack Obama, to be the President of America! That was a good start, but more actions in this direction need to be done. Did that event stop any of the repulsive treatments black people receive? The most disturbing part of the expectation is the demand for gratitude from the black people because they were brought out of Africa's 'hostile' environments and into civilization and made Christians. The violent and brutal treatment of the unfortunate Africans and the loss and deprivation of their homeland, Africa, can't be termed civilizing.

Suppose a 'civilized' person/police officer shoots and kills a black man. In the 'court' of law, the officer only needs to claim that immediately he saw the black man, he feared for his life. The 'civilized' Jury will conclude that the police officer has a right to protect himself and most often return acquittal. What if a black man running away from a police officer? Note, he was running away from and not attacking the police officer! Our law has it that the officer has the 'moral and legal' rights to shoot and kill the 'criminal.' You can see justice in action!

How do other societies see America? How do they feel about such actions, particularly if it is on a regular, not isolated occurrence in our communities? But we ask, "why do they hate us?". What about the commandment: "Thou shall not kill"? Please think of how many innocent people are killed every day on our streets!

A white friend of mine related an encounter between his father and a black boy. The kid entered the father's shop and stole a twenty-five-cent candy bar. On hearing some noise in his shop, the father appeared with his shotgun in his hand. On seeing the boy running away, he took a shot and killed him! On examination, neighbors attracted by the firing discovered the boy had a twenty-five-cent candy bar in his hand! The shop owner did not have to be arrested. Why? It was self-defense. Might you think that was an isolated case? But, No!

Those unfortunate Africans were stripped of anything that made a person, a human being: family, loved ones, name, custom, and above all were inundated with the lies that they were so bad, so disliked by their families, they sold them. In this state of mind, these Africans were required to show they were pleased people who adored the 'redeemer'/masters but to hate Africa or whatever would remind them of or connect them to Africa.

When will the anger over this type of treatment simmer and end? How will that begin?

Should the anger disappear because the original perpetrators are no more with us? The burden of reparation and amendment rests on the descendants. Yes, reparation – a show of remorse. That is the only way that makes sense.

Unfortunately, few of these descendants agree that "I didn't do it," but the mere fact that I enjoy or profit from the evils" should make them feel some obligation. But some do not accept any responsibilities. By being spoiled by years of wealth and privilege, some behave like they would have done the same if they had lived in that generation. Is it true we don't know "why they hate us"?

Many Africans and African Americans have the most forgiving souls. Otherwise, how could they be asking only to be treated as human beings with civil rights in a country that their sweat helped to build?

A lot of improvements have taken place in the lives of minorities in America. A lot more is required. Time for anger and disillusionment over past ill-treatments is over or should be put behind us all. All the people concerned should remember the sacrifices made by many white and black Americans to secure justice for all and try to use whatever opportunity available to them to improve their lots. The privileged white people should try to wear the shoes of the African Americans and other minorities to understand what these people are going through. It is possible if they have the Will.

The use of the word "They" or "We" should be revisited. Both Black and white Americans are all Americans. America has room for all its citizens. All we need to do is to look around us to see God in everything that He created. All of us are passing through this earth. And in the end, we will leave the planet as we entered it – alone and with nothing.

The IRA

Another Brand of Freedom Fighters' Organization is the Irish Liberation Army (IRA), formed in 1919. The main objectives of the organizers were:

- End British Rule in Northern Ireland.
- Unify north and south.
- Finally, gain Independence for the whole country of Ireland.

Again, this is a rebellion against a mighty Country with mighty force. All their operational tactics were the same as all Freedom Fighters: sabotage, disruption of the everyday life of the citizenry, hunger strikes, marches, and boycotts. Their weapons are also the same: explosives or bombs, firearms, raids and kidnappings, assassinations, and guerrilla warfare. The armed struggle is over; some of the objectives are yet to be realized.

Ref.: Encyclopedia Britannica (www.brittanica.com/topic/irish-republican-army.

Imagine how many families lost loved ones to armed operations by these 'terrorists' or by the British reprisals? This man-generated unrest and misery were caused by the British imposition of its rule in the Northern part of Ireland. Britain is still in that part of the country today. Ireland is not part of Britain. Or is it? There is no justified reason why Britain should be there.

If you did not hear or see, you could imagine the amount of unhappiness generated by the violent actions of the freedom fighters, or should we call them 'terrorists' for asking Britain to leave what belonged to them? Who can believe it! The whole world body has stood by and watched the atrocities going on there for decades? Who can tell Britain what to do, a country once the greatest power on earth?

Those who sell arms to make money or, as they put it, 'help a friend' are not really interested in seeing the instability end. Those people are only interested in their bottom line. Is there no religious leader, president, Prime Minister, or any group of these important people that could persuade Britain to do the right thing? Or did Britain develop sudden amnesia over the laws of the Creator, which she, Britain, propagated all over the world – not to usurp what belongs to another, not to kill?

When I read about the horrific evils British perpetrated around the world, I began to wonder about the opinion I had about them in the past. Could it have been childishness? I thought they were people without sin - perfect creations of God. Those people were above the pettiness of life. They were always concerned about the good of the people they encountered. They sowed the seeds of the "Good News," the acceptance of which was a sure way to 'heaven' and the rejection leads straight to 'Hell.' They taught their converts always to remember the day of judgment.

I came to understand that there was hardly any country they colonized that escaped poverty and the never-ending conflicts. Would the actions of Britain in Ireland, in Nigeria, India, or any of their colonies, not affect and shape the thinking of current and future generations of those peoples?

But the British will be the first to shout, 'Terrorism,' 'Bad people who hate our way of life' when an Irish or any of the colonized peoples takes up arms in retaliation for what he or his people suffered in their hands. Britain can't claim it does not understand how people feel when they are dispossessed of their things, enslaved, killed, or jailed for fighting for their rights.

There is no doubt that the religion they brought with them was the main instrument that helped them gain access to the indigenes and kept them in line. It did not matter the country. The effect was the same as described in this book. Where do we go from here?

It is easy to see many regrets from the current generation of the British and their resentments against past behaviors. Besides, during the occupation of those regions of the world, not all the British supported and participated in the atrocities. Many of those in opposition were sent to jails or even killed.

Although there is a lot of resentment against the part religion played in all these, there was a lot of good that came from it – the establishment of schools, medical facilities, the abolishment of certain cultural behaviors like the killing of twins, human sacrifices, the end of slavery, etc. Granted, all cultures had their negatives, which in time would be corrected, the part they played to end those practices can't be ignored.

There is no way any person can go back in time to undo what was done to these people of the world. But a public acknowledgment and apology would help – the Creator's expectations.

PLO

From the Middle East comes the case of the Palestine Liberation Organization (PLO), established in 1964 during the Arab Summit in Cairo. This is an armed struggle against the Israeli Government with the sole aim of the liberation of Palestine and restoration of the Palestinian homeland. Their violence was directed against the Israeli Civilian Population.

The PLO is recognized as the legitimate representative of Palestinian people but labeled a terrorist organization by the US and Israel. This characterization changed in 1993 after PLO accepted UN Security Council Resolutions 242 and 338 and rejected violence and terrorism.

A PLO splinter organization, Hamas, still bears the terrorist label because it refused to renounce terrorism and, above all, to acknowledge the existence of Israel as a nation.

Before going further, it is necessary to bring the State of Israel into the discussion.

The State of Israel

The State of Israel cannot be left out when the unrest and instability in the Middle East are discussed. In a nutshell, some of the events of 1917 were: The British took over the rule of Palestine, and the Balfour Declaration of 1917 paved the way for the caving out of a part of Palestine to be the home of the Jewish people. From 1919 began the mass exodus of Jewish people from all over the world to their new home in Palestine.

The Palestinians and other Arab people refused to accept the partition, which eventually gave rise to Palestinian Nationalism. By 1938, more than a half million Jews had arrived in Israel. The return-home fever increased in tempo during and after the Holocaust and the end of World War II. On May 14, 1948, the State of Israel was declared with Ben- Gurion as the first Prime Minister and immediately recognized by US President H.S. Truman and USSR Leader J. Stalin.

On its Independence Day, Israel was attacked by combined forces of Palestine, Egypt, Lebanon, Syria, and Iraq. Israeli forces defeated these attackers. The victories resulted from excellent preparation Israel made toward Nationhood with the awareness that it had hostile neighbors. By 1948, Israel had trained soldiers that reached more than one hundred thousand in number by Independence Day. After routing her enemies, Israel began a massive expansion program beyond what it had been allowed by the United Nations. The resulting tensions culminated in the 1956, 1967, and 1973 wars.

Palestinians did not agree that Israel was the legitimate owner of the land they were occupying. And the subsequent expansion by Israel did not help matters. Israel's claim to the land dates back to the 'biblical promise' God made to Abraham.

It is instructive to mention here that Israelis and Palestinians descended from the two sons of Abraham from which Israelis claim ownership of the land. It is also noteworthy to point out that the Israelis were occupying that land right from the beginning until driven out and scattered across the globe for more than a thousand years. During this time, descendants of semi-nomadic Bedouins, Arabs, Babylonians, and other Arab groups took over and occupied the land. This occupation is the crux of the conflict – i.e., who is the rightful owner of that piece of land?

It has been explained elsewhere in this book that by examining the behaviors of the various creations, it is easy to observe that the Creator made self-preservation the first duty of all living things. Understandably one must defend oneself when threatened, including using deadly force (preferably only when necessary and justified).

We need to be careful here! It should not be understood that self-preservation means selfishness. We must 'live and let live.' That's the Creator's law. A cow or any other animal should be killed only for food, or if left alive, will bring sickness, danger, or death to others. We are given the brains, senses, and conscience to control those impulses. Even as small children, we were taught to value and respect life, i.e., 'don't kill any living thing just for the fun of it.' Then think of taking another person's life! Killing is an act that requires careful consideration, performed only when self-preservation, defense/protection of human lives, or a penalty for murder (in some cases) requires it.

The Creator placed humans at the highest level of all His earthly creations by the special endowment of the brains—intelligence and the conscience. With this unique gift, man can perform his divine-assigned duties, including taking care of His earth. But men make wars or perform acts that can cause death. Generally, animals kill only for food as the Creator requires of them. But humans don't kill only for food or to maintain stability or safety as the Creator commanded. Some other reasons men have killed are greed, to show power, influence, or authority. We know the consequences for disobeying the Creator's laws – the misery and unhappiness that result from those actions.

Are there just wars? Was the war to get rid of Hitler and his henchmen justified? What of the one waged to get Iraq out of Kuwait or assassinate Saddam Hussein or Muammar Gadhafi? What of the wars Israel waged against some Arab countries when those countries besieged it on its Independence Day? What we think is the correct answer does not matter but what the Creator requires of us - the need for each other and live as cohabitants of the earth.

If both warring parties – Israel and her Arab neighbors, accepted that it is the Creator who owns all lands and has allowed us to occupy a portion of it during our brief time here on earth, the fighting and killing would not have occurred. The Jews, Israelis, and Arabs have a right to and must occupy a place in God's lands.

But the rebellious humans will always think they would do better than the Creator. None of the parties, Israelis or the Palestinians, obeys the Creator's natural law to "Live and let live," explaining why the Arab Nations could not accept the U.N.'s carving out the Israeli land from Palestine.

If the blinding effects of greed and ego are excluded in the conflict, it will be seen that the Creator made the lands spacious enough to accommodate all His children indeed.

Supported by powerful countries, like America, France, etc., that supplied all the arms necessary to fight its enemies, Israel embarked on an unnecessary expansion, unconcerned about the sufferings of those displaced or exiled by the UN mandate and the wars. We are all witnessing misery brought about by selfishness and ego.

STATISTIC	ISRAEL	PALESTINIAN	
Children Killed	134	2,154	Online a@if americaknew.org
Adults Killed	1,213	9,478	Online a@if americaknew.org
Injured	11,838	92,452	Online a@if americaknew.org
Military Aid From USA 2016	$9.8 millio/Day	$0 Per/Day	The Congressional Research Service Report: US Aid to Israel
UN Targeting/Sanctioning	77	1	Online a@if americaknew.org
Political Prisoners and Detainees	0	6,500	Online a@if americaknew.org
Demolition of Homes since 1967	0	48,488	Online a@if americaknew.org
Unemployment Rates	5.6%	West Bank 17.1% Gaza 44%	CIA WORLD FACTS: - Israel – 2015 - West Bank – 2014 - Gaza Strip - 2014
Illegal settlements (on other's lands)	261	0	America for Peace Now (B'Tselem)

Fig. 9: A Particular Statistics on Israel/Palestinian Conflict.

The statistics speak for themselves. It requires the hard-to-come-by courage for all concerned to look around to see the handwriting of the Creator displayed on all His creations, boldly proclaiming the universal brother/sisterhood. "We must be our brother's keeper." The rest of us are too removed from the scene of conflict and therefore less likely to appreciate the weight of the situation from the Israeli perspective. Only very few of us have been driven from our lands and scattered all over the earth for decades to endure the humiliations of strangers in alien lands or see their need for a place to call home.

Neither can we, particularly those more interested in personal gains, see the situation through the lens of the Palestinians (the Hamas). It was not our sisters, brothers, mothers, fathers, and babies that lost their lives in the conflict or have shelters as permanent homes. No! It was not our houses, buildings, businesses, and homes that were destroyed in the fighting. Many of us saw these played out over our TVs. Likely, they did not disturb or interrupt our daily routines.

Israelis have made a lot of concessions resulting in a lot of compromises. But more such moves remain to rain in Hamas. A steady foot must be maintained in this quest for peace, remembering that we are simply 'passing through' this earth. All we have accumulated and call ours belong to the Creator. We will leave this earth as we came – alone and with nothing. Palestinians and Israelis need peace most, including a good night's sleep.

Here we are faced with, supposedly, a conflict that defies a solution. Its basis is an ancient family feud in which the world has taken sides to serve their interests. How can the US, for example, negotiate for impartial peace if it, at the start of the negotiation, declares unconditional support for Israel no matter what! This attitude is the main reason peace has been so elusive.

Israel has managed to make peace with the rest of her hostile neighbors except Palestinians, who have the PLO as its representative, and Hamas, a splintered arm. The PLO has agreed to end hostilities, renounce violence, and recognize the existence of Israel as a Nation. However, Hamas rejected all forms of peace and would not accept the existence of Israel and continue acts of violence and terrorism to date.

Yes! Israelis engaged in those wars for self-preservation. What would any country do when besieged by hostile neighbors? More work remains to be done while Israel has made concessions like withdrawing or ready to withdraw from lands (booty) captured during the wars. Might Israelis, particularly those who once held rallies against Israel's disproportionate use of force, redouble their efforts, and all parties remember that everything belongs to the Creator. In the end, both will leave the earth as they came – alone and with nothing.

The statistics displayed above bear witness. How can Israel and its supporters forget that as their children are sources of joy, happiness, and a blessing to them and the world, the Palestinians derive similar pleasure and feelings from their own? How can they explain the deaths of those Palestinian babies to their children? What of the thousands of houses and buildings demolished?

Most of the houses belonged to the poor. Would self-preservation explain to the children and the next generation of the people in that region that the destruction was unavoidable? What of the thousands of Palestinians and Israelis languishing in refugee camps for decades? Israelis, of all people, know refugee status firsthand. Doesn't Israel have the wherewithal to bring Hamas to the bargaining table?

Indeed, Israeli people would sleep a lot easier at night with a complete cessation of all hostilities and would be thankful to those who claim that the unfortunate brutal acts were committed all on their behalf. Regional peace will serve national interests better than the bottom line of friends, families, and those who negotiate for or supply the instruments of death.

How much will, say, five warplanes, five armored cars, and five tanks cost? Or are they supplied free of charge with no strings attached? The money spent on those small number of war implements is more than enough to bring Hamas to the bargaining table. This is possible and should be done; considering that Hamas and their people are the owners of the lands now under occupation by the force of arms; given by UN mandate; whose houses were demolished and their loved ones, dead or in refugee camps never to return, Israel is known to be the master of precision warfare and knows how to go straight to its target without unnecessarily killing and destruction. Remember what happened on its Independence Day in 1948, when it single-handedly defeated the combined military forces of five countries: Palestine, Egypt, Lebanon, Syria, and Iraq. Or in 1967, war, now famously known as "The Six-Day War," where it defeated Jordan, captured the West Bank, defeated Egypt (capturing the Gaza Strip and the Sinai Peninsula), and defeated Syria (capturing Golan Heights). It was the Israeli commandos who, in 1976, mounted the most daring rescue operation in Entebbe, Uganda, freeing 102 out of 106 Israeli hostages held by PLO Guerrillas.

But war, no matter who initiates it, has deadly collateral consequences such as the deaths of hundreds of Palestinian children. Everything possible must be done not to repeat this.

Indeed, Israel had strong backing and sophisticated military hardware to wage these wars. But Semitic people will be happier with a negotiated peace than lands acquired by force of arms, resulting in mass killings, destruction of houses, and untold sufferings. Remember, at the end of it all, all earthly acquisitions will mean nothing!

We have certainly been taught wrongly by our religious instructors about who the Creator really is? Certainly, agents of war and destruction don't understand the power of the Creator and His law to be "your brothers' keeper" and the consequences for failure or disobedience? Humans need to take time out to go back to nature to learn God's laws and their duties here on earth and then receive the Creator's Blessing of happiness. There is no other way.

Adapted from: (ifamericaknew.org. / Published by Jew for Justice Berkeley CA 2001 / The Origin of the Palestine-Israel Conflict).

Al-Qaeda

The world's most notorious terrorist organization was Al-Qaeda. In a nutshell, this movement was founded in 1988 by Osama bin Laden, Abdullah Azzam, and other Arab volunteers who fought against the Soviet invasion of Afghanistan in the 1980s. It was headquartered in Afghanistan, with training camps in countries all over the world.

Unlike regular freedom fighters who restricted themselves to their own countries, al-Qaeda mounted attacks on both military and civilian targets worldwide. It was responsible for, among others, the 1998 US Embassy bombing, the Sep. 11 US Twin Towers attack, the 2002 Mali bombing, and many others, including targets they considered Kafir.

Grievances:

They believed that the Christian-Jewish alliance was conspiring to destroy Islam.

They opposed human-made laws and wanted to replace them with strict Sharia laws according to the religious precepts of the Quran and Hadith.

They believed that the killing of non-Muslims is religiously sanctioned.

Al-Qaeda leaders regarded liberal Muslims, Shias, Sufis, and other sects as heretics and had attacked their Mosques.

Their leaders envisioned a complete break from all foreign/outside influences. The members pledged loyalty to their leader, Bin Laden, or any of the affiliates who had trained in one of their camps.

Due to the intensity and number of attacks mounted against them by the Coalition Forces, Al-Qaeda did not enjoy a haven anywhere in the world. Al Qaeda has lost most of its credibility because of its attacks on other Muslims and the death of its leader, Bin Laden, under the leadership of a US President, Barack Obama.

Ref.: Brookings Institute/Wikipedia – (https://en.wikipedia.org.org/wiki/al-Qaeda)

ISIS

The most notorious terrorist organization was the Islamic State of Iraq and Levant (ISIL), also known as The Islamic State of Iraq and Syria (ISIS).

This group was a terrorist organization founded on the Wahhabi doctrine of Sunni Islam. It prided itself on the global Jihadist principle and followed the hardline ideology of Al- Qaeda, with the de-facto capital, Raqqa, in Syria. Almost all ISIS leaders were the former Saddam Hussein's Ba'ath party soldiers. It came to world recognition in 2014 when it drove Iraqi government forces out of and occupied key Iraqi western cities. This victory was followed by its capture of Mosul and then the Sinjar Massacre. Its record on human rights brutality and other war crimes was bone-chilling.

Starting as an arm of Al-Qaeda, ISIS quickly proclaimed itself a Caliphate following the 2003 invasion of Iraq by Western Forces. On June 24th, the following year, it began to refer to itself as the Islamic State, a term that generated a lot of debate and controversy throughout the Muslim World. Abu Bakr al-Baghdadi (later, killed himself on October 27, 2019, by detonating his suicide vest when under the crosshair of the US army) was proclaimed the Supreme Leader, the Caliphate, a man they claimed to have traced the lineage back to Prophet Mohammad.

It began its expansion program, and by 2015, it had conquered a large area in Iraq and Eastern Syria, where they enforced its interpretation of Sharia. It was well situated in terms of finance, human resources, and effective organization. Because of their recruiting methods and ideology, young men and women flooded to them in droves from all over the world. Their brutality includes beheading.

With the Caliphate destroyed in 2019, under US President Donald J. Trump, ISIS operates virtually in more than eighteen countries.

Ref.: (http://en.wikipedia.org/wiki/islamic-state of Iraq the levant).

These two world-recognized terror organizations must be considered together since both are diametrically the same. Where will anybody begin to describe any organization whose philosophy and brutality defy all reason? (Or are they?). And questions our understanding of why young men and women with backgrounds of varying religious persuasions flooded to them from across the globe. What can induce any person to agree to wrap around a suicide belt, bombs, explosives, etc., and be prepared to detonate them to kill as many people as possible, including themselves? What will make a person fly an aircraft carrying hundreds of people into a building, intent on killing everyone and himself?

Some people see these two organizations as a menace and believe the two can be destroyed simply by ‹removing the head› and leaving the rest of the body to die on its own. Thus, their leaders have been targeted and eliminated. Also, their financial and cyber backings had been broken. When this strategy failed to produce the intended results, the coalition governments decided to take the war to them. This strategy has been more successful, delivering a mortal injury and slowing them down almost to a halt. ISIS has not been wiped out completely. Remnants and pockets acting as lone wolves remain and active in many parts of the world.

I think, ultimately, we must take time out to examine what motivates and sustains a person who is prepared to kill as many people as possible – people they had no dealings with, never seen before, or had any quarrels with in the past. Above all, he is prepared to sacrifice his own life doing it. Some theorize that these young men and women are drawn into those brutal actions by the lore of monitory rewards or the promise of the 'Key of Heaven.' Could any of these be true?

We all know the Creator's laws on self-preservation, usurping what belongs to another, and the consequences for contravening any of those laws. We are also aware of the manifestoes of these two terrorist organizations.

All I got from reading the Documents could be summarized in one sentence: "Don't interfere in their culture – their religion or their way of life." They were avenging what they saw as injustices done to them. Are they? Reasonable and fair-minded people of the world, good Christians, Muslims, Hindus, and adherents of all other religions, it is incumbent on us to find a solution to this evil called terrorism in all its forms.

The outcomes of human interactions are complex. Therefore, we must be careful not to paint everybody with the same brush. I'm referring to the misuse of the words "they" or "we" in statements like, "they" did . . .. to us and "we" suffered . . ..

We know how the world came together to fight the evil represented by Hitler and his henchmen. There was no justifiable reason for Hitler to conduct his offensive campaign on the Jewish people. Imagine how many Germans were imprisoned or even killed because they tried to save the lives of some Jewish families by hiding them in their houses or caught smuggling food to them. I call these Germans uncelebrated heroes. Would it be fair to accuse all Germans of supporting Hitler?

Through the same lens, I look at the evil called the Transatlantic Slave Trade. Some God-fearing British and American citizens spearheaded the fight against and ultimately secured the abolition of the slave trade. Again, there was no justifiable reason for the inhumanities of slavery to take place. It is not fair to paint all British, Europeans, and Americans with the same brush of guilt.

The Creator made each human (each living thing) on this earth a unique individual, gave him the brain and the senses to be able to determine his own destiny and the conscience to control his actions. So, every person is responsible for his own actions.

The same story can be told of slavery in America. That will call to mind the bravery shown by many white Americans like the Quakers in the Underground Railroad, including their participation in many protest matches and providing financial support for Slaves' legal battles. Many white people were killed trying to help the cause of the freedom fighters. When the evil called Apartheid reared its ugly head in South Africa, fair-minded people fought and stamped it out because the British had no justifiable reason to occupy South Africa or enslave the people.

I discussed the actions of some freedom fighters from different parts of the world earlier. A common denominator was evident – all fought to redress evils done to them, specifically, the imposition of alien authority on their people and usurping the people's natural resources and, in most cases, oppressing or enslaving them. Yes, in these cases, the evils had come to an end, mainly due to their resistance and the heroic actions of many people.

Was it justified to conduct the 9-11 massacre or the movie theater shootings or stabbing innocent people or mowing down people using trucks because "they" (or their people) did evil against "us" and "we" suffered a great injustice? Should those that received the injustices remain angry and hateful to the point of shedding innocent blood? You make the judgment. I think it is best to seek the Creator's intentions in all.

In fairness, ISIS and its senior brother, Al-Qaeda, must be subjected to the same scrutiny as other freedom fighters. Despite their disdainful methodologies, we need to look at them through the same lens to try to understand their motivations, albeit misguided methods.

After creation, God, in His infinite wisdom, placed humans and all other living things in different areas of the earth. He created the human 'machines' and knew what the machines needed to perform their tasks here on earth. To aid them to complete their assignments and be the masters of their own destinies, He provided them with the brains, senses, and the conscience as their guard.

All the dark peoples of the world, for example, He placed at the equatorial regions of the earth. These peoples, by His prompting, developed their ways of life. We can't develop sudden amnesia as to why He did not endow one person or a few groups of people with all knowledge.

The ability to develop knowledge was given to every living thing, but each received according to its nature and purpose for which it was created. These are universal truths.

How humans have treated one another over the years and respected the laws of the Creator have been discussed (in detail) in this book. They were found to be less than desired, the root causes of all the unhappiness here on earth. To provide context, I will examine a few contemporary events.

In no distant past, there was a movement by the West headed by the US under President George W. Bush to spread democracy all over the Middle East. The West was particularly irritated by the way Muslim women were 'treated over there.' In other words, 'Feminism,' for example, was not allowed to thrive there as in America or the West. We still remember the Arab Spring that started in Egypt and spread to other Middle East countries. Then came the massacre of 9-11, a single act of revenge (if we have the moral strength to call it what it was) that forced about three thousand innocent peoples to their early graves.

Al-Qaeda, led by Osama Bin-Laden, was identified as the culprit and had taken refuge in the mountains of Afghanistan. The family of bin-Laden was safely evacuated from the US to Saudi Arabia, their home country. Let's not forget that during the Gulf War previously, to force out Iraq from Kuwait, the coalition forces were quartered in the 'Holy' land of Saudi Arabia, which infuriated many Arab Moslems.

Then began the war in Afghanistan to get Bin-Laden and his henchmen to pay for their crimes. Suddenly, the war was expanded in scope to include getting rid of the Iraqi President, Saddam Hussein, because "he possessed weapons of mass destruction, WMD and had killed his people." Finally, Saddam and some members of his family were killed. Was WMD found in Iraq? No! How many children, women, and men were murdered because Saddam killed one hundred of his people (for whatever reason)? The number is in thousands.

Then ISIS was born comprised of members of Saddam's special guard. These were the "bad guys" while we, who killed thousands of people there, destroyed properties and businesses worth millions of dollars, are the "good guys." These wastes did not include billions of dollars that the coalition countries wasted funding the war, with thousands of coalition young men and women killed or returned home with steel arms and legs.

Compare this to the colonization of the third world by the Europeans championed by Britain. Do we have the moral strength to call a spade a spade? We know stepping outside the Creator's laws will always incur consequences. Who was surprised that some Western sons and daughters, on their own, joined ISIS or became home-grown terrorists? This act alone by these young people should tell us we were wrong. The Creator was not happy with our actions, and by the Creator's designs, the chicken came home to roost!

Killing, not to talk of mass-killing of people (or any living things), is forbidden by the Creator, except in some extreme cases like those mentioned earlier. Hardly can any person point to the 'extreme' circumstances making it justifiable for these terrorist organizations to carry out mass, brutal, and heartless killings. Do we see the "extreme" conditions? What of the claims that their God-given rights had been infringed upon; what belonged to them usurped; that they had no other way to avenge a mightier power or defend what belonged to them?

Seriously, can we look ourselves in the face and deny that we have not attacked their religion, tried to influence, or change their cultures and manipulate them in different ways for our gains? For starters, consider what happened in Iraq – the number of children, women, and men we killed to get one man. That is the answer! Do we think that these people are senseless animals that have no feelings or don't understand when their rights are taken away?

Without more explanation, we all agree that the culture of people is who they are and the Creator-sanctioned. And they have the right to protect it. The fact that our own sons and daughters joined them (as mentioned before) to fight against us for their rights is a phenomenon to which we must give a serious thought.

Consider the West's exploits all over the earth, particularly in the 'third world' countries. Yes, third world! These exploits by the West were carried out in such a way that the permanency of the ensuing disasters remained 'forever.' Why? Why would any human being treat another that way? All the colonized countries suffer a similar fate: besieged by never-ending wars, insecurity, poverty, antiquated or non-existent infrastructure, old health care system, name it?

Consider the Berlin Conference of 1885, in which the Continent of Africa was scrambled for and divided up by European countries. All for procuring raw materials and nothing to do with the welfare of the indigenes. No representatives from those African countries were invited to the conference. With little credit to our history education and much to the internet, we all can access this information by simply googling it. It is better not to repeat the type of brutality the foreign countries imposed on Africans and all colonized countries to secure free raw materials, free/cheap labor, etc. Again, why?

The British have horrible stories to tell their grandchildren about exploits in West Africa and other countries they occupied throughout the world; the French have Congo on their list of countries that suffered humiliations and brutality in the hands of their fore-parents. Don't forget the brutality of Spain, Portugal, Italy, etc., on innocent African peoples. Did those simple-minded Africans invite the Europeans for any reason? No! The colonizers moved in by the force of the Bible and arms.

The Creator-forbidden slave trade and the horrors of slavery will, forever, remain a scourge on the conscience of humanity. The subhuman treatments meted out to many Africans in America, even to date, are there for all to see as evidence of hypocrisy by people who go around preaching human rights, godliness, and the evils of killing. Many times, perpetrators of these evils misunderstand that people are keeping tabs on their misdeeds.

I spent considerable time in Ghana, precisely in Cape Coast, between 2009 and 2011. I had the opportunity to visit the two slave castles – one in Cape Coast and the other in Elmina, both almost within sight of each other from the sea. It was shocking to still smell the odor of rotten human flesh on the walls of the castle, even to date, after more than three centuries. Imagine what happened there!

Two other things caught my attention in the castles. One was an opening cut in front of the balcony of the top floor of the massive structure where the Governor lived, to the holding compartment for the African women slaves in the dungeon below. The Ghanaian tour guard explained to us that the opening provided a clear view for the Governor to see and select a choice woman, any time, for his pleasure. He ordered that his choice be cleaned up and escorted upstairs for the master's fun.

The second observation was that while slaves were kept in the dungeon below, suffering the agony of their misfortune, church services were conducted on the top floor. Church service! Amazing!

A general argument now has been that all African countries gained independence many decades ago. This leads to the question of why no change for the better but a steady decline in every sphere of African people's lives? This question has no place in this discourse.

Take Nigeria, for example, but without going into detail. Nigeria has a large population, one of the largest in the whole continent of Africa. The Creator blesses it with many arable, fertile lands and perfect weather for agricultural activities. Above all, Nigeria has crude oil and other natural resources.

Since no one can point to any meaningful development in the country from the abundance of natural resources, one is inclined to ask what happened to all the money realized from the resources. The money went to mass-produce millionaires and billionaires. But the Nigerian treasury was left empty? Ok! The Millionaires and Billionaires must have been using their loots to build Industries and carrying out a lot of development in the country? Continue dreaming! All the money was siphoned away to foreign banks to enrich the countries supposedly 'helping' Nigeria. The chaos resulting from this gave rise to what is known as the 'Brain Drain,' a condition that forced high IQ Africans and geniuses to leave their countries to work for and develop other countries while the resource-rich continent collapses. Beautiful story, right?

My case here is an appeal to all concerned, good people of our world. If somebody pinches you, yes, you will feel the pain. Then, do not forget that when you pinch somebody, the person will feel the pain likewise. Middle Easterners, Africans, and the rest of the third world have every right to grieve over injustices done to them. All humanity should weep for them too. It will never be productive to permanently maintain silence or bitterness for evils done many years in the past. Most of the perpetrators have already left this world.

Current generations must come together to calm anger and provide redress. Pointing accusing fingers on one hand or not acknowledging the pains we were indirectly or innocently involved in is not the Creator's way. We must work together on the solutions and remember that we have our responsibilities for the future generations and how we conduct our affairs here on earth as prescribed by the Designer. Let's not forget that these earthly accumulations will be of no consequence at the end of it all.

Some people might disregard doing the right thing by these hurt and injured people, believing that the force of arms, trickery, and hypocrisy will calm the aggrieved and dispossessed. So far, those strategies have not worked and will never work. The countries

(the inheritors) that inflicted these atrocities on innocent people know the just and right thing to do for brutalized people.

Chapter 22

CONCLUSION

I'm sure that immediately you picked up this book to read, you could not help but wonder about the title, Time Out. How do you see it now: strange, appropriate, or food for thought?

Some people might find this book very restrictive, preferring to read about a merciful God that will forgive any sin if they ask for forgiveness. I bet those people did not read or understand God's laws as written in his creations around us. And so, they forgot that the Creator is the supreme lord of justice. He can't look the other way because, for example, one prays five times a day or is 'washed' by the blood of Jesus, etc. Rewards and consequences are already built into His system, and no human or anything can influence or change His decisions.

Time Out is not telling anybody what to do or making decisions for anybody. If that is necessary, the Creator himself would be telling or guiding people to what to do. Instead, He gave humans free will, intellect, senses, conscience, etc. This book is merely pointing out what is in front of us humans, the laws of God boldly written in His creations in our environment.

I requested for you to read the book with an open mind. Now you have gone through the book, we may be on the same page, i.e., that our universe was designed by one all-knowing and all-powerful spirit we call **God** in the English language but different names in other languages. We know He created our earth, one of the planets of this universe, and made it the only known home for humans and all living things. This is true because He created the earth complete with everything that the living things would need to sustain life and complete their tasks.

We also know that because of the interdependency among all living things and their abilities to reproduce themselves, He intends for His creatures to occupy this earth to perpetuity and happily. On human beings only, He bestowed the gifts of intellect and free will. It is reasonable to believe that He issued (etched in their DNA) His commandments to all living things: "You must go for your food, and you must eat the food to stay alive. I designed your body and the type of food suitable for it.

To get the food I have already provided, I have given you the means to do so. Any attempt to alter my designs will incur consequences. You must protect yourself and what is yours from predators." To humans only, He gave the charge of taking care of His earth, using the unique gifts of their brains, senses, conscience, etc. Stepping outside these laws indeed incurs consequences.

The central theme of **Time Out** is: How did all the evil things we witness daily enter our otherwise beautiful earth? Two contexts are examined, nature and nurture. Nurture was found to be the main culprit: including the failure of humans to understand who their Creator really is, why we must heed His laws, our purpose for being created and put on earth, and the lack of appreciation of the Creator's outliers. All these are sources of unhappiness here on earth. It is shown that **time** is required to transform the earth back into the original design and intent of the Maker – a Paradise.

There are different schools of thought as to why God created the planet Earth, furnished it with everything that human beings would need to live on it, and then put human beings on it. A group believes the purpose was for human beings to worship, serve, praise, make sacrifices, and glorify the Creator. Could this be true? The answer is 'no.' Why? God is a spirit and does not need earthly things. Spirits can't eat or drink or make use of any earthly materials.

The praying, the worshipping, etc., that humans do is natural to them. When a young one – animal or human is in pain or need-simply goes to the parent. Even adults do the same thing in similar situations – they go to whoever they believe will solve the problem. There are so many situations that defy human solutions. Then they seek a divine solution. This is how going or praying to the Creator came to be.

A reason offered by another group was that human beings were created and put here to take care of the earth. It is easy to believe that human beings take care of the earth, but does the earth need to be cared for by humans? No! If it were the will of the Creator, the earth would not need the care of humans – it could take care of itself just like other planets – Mercury, Mar, Saturn, etc.

The final (educated) guess was that the real reason humans were put here on earth, or the universe created was because the Creator so desired it; nobody knows! That's how far the human mind can go. How can the little and primitive minds of human beings interpret or understand the thinking of the Almighty God? What we do know is that all humans want to: survive. And be free to continue enjoying all the good things that the Creator provided for all His creatures.

Since humans can't see God or communicate with Him, the time has come when humans should open their eyes to see His laws boldly written in His creations all around us. They should then embrace the laws and obey them to end all the unhappiness resulting from disobedience and enjoy all the good things God has showered on His creatures.

Finally, is it not surprising that human beings chose not to obey the Creator that created them, resulting in all the unhappiness that has become the way of life here on earth? Why? Do humans really understand that the Creator is God, who provided all they need to live here forever and happily? The answer is no! If they do, they would obey His laws and end all suffering on our earth. The following will explain why my answer is in the negative:

One day I observed my friend receive a phone call from his boss. The boss was holidaying in London, about four thousand miles away. The boss called to inquire about how things were going with my friend's charge. What shocked me was that immediately my friend said, 'Hello,' he fell on his knees. I waited patiently to the end of the call before I enquired as to the reason for his reaction to a simple phone call. He replied, "You can't understand."

Then he started telling me that it was the boss' help that had kept his family from falling into poverty and the reason two of his children were attending college; every time he got home from work, there was food on the table and his wife a happy woman. He never expected so much from just a friend. He concluded by telling me there was nothing he could not do for the man, including risking his life for him. I was dumbfounded.

My thought sprinted to my parents› behavior towards the Creator before foreign religions took root in Nigeria. My people, who could not even alter the name of the Creator out of respect and fear to displease Him, were called 'heathens' by people who continuously disobeyed the laws of the Creator. My people understood that God gave them everything, including their lives. Then I concluded it was due to this relationship – complete understanding of who the Creator really is and total dependence on Him that any hint of disrespect was unthinkable among my people.

But my friend and his boss are humans, mortals. Unlike the Creator, humans have limited knowledge of one another, including their thinking. One person can't know and satisfy all the needs of another person. But God is a spirit who created us and all things.

All we need was known to Him before we were brought into existence. It is clear we can't even begin to compare the knowledge of the Creator of His creations and that of my friend's boss of him.

The level of respect and honor accorded to any person depends on the functions or the importance of the person. In any healthy family, the respect accorded to the father and mother depends on their usefulness to the family. If a President, Prime Minister, Queen, or King of a country is visiting one of the cities or even another country, preparations to receive him will be very high to show respect to the high office. But a father or a mother will attract much less attention doing the same thing. If the Pope of the Catholic Church issues a 'decree,' the order is carried out immediately simply because of his high office.

But humans defy the laws of the Creator with careless abandon, using excuses like: "God loves me, He will always forgive me or look the other way"; "My God is kind, not wicked." "I only need to confess my sins"; "If I could secure the key of Heaven, Paradise is assured no matter what evils I commit"; "The blood of Jesus washes me"; "all I need to do is pray five times a day"; "If I die before the age forty, Paradise is assured"; etc.

Despite these beliefs, humans who disobey the natural laws of God continue to incur consequences. What might be the reasons for their adherence to these false beliefs but find it difficult to obey God's laws? Even Jesus warned that he would not know those who "did not do the will of his Father, God." The only reasonable explanations are they don't know and care to know who the Creator really is. It does not stand to reason why humans will readily obey the commands of officials who are merely limited beings like themselves but find it difficult to respect and obey the Creator's laws. Disobedience to the Originator of life, the Creator of all their needs and comforts, will always incur retribution – all the unhappiness here on earth. What makes it possible for humans to continue to exist here on earth and enjoy all the free bounties of the Creator, despite all the human disrespect to Him, is simply that it is His will, not because we humans deserve it.

The teachings about the Creator obviously were not founded on absolute facts and truth. Otherwise, humans would obey Him. His creations or the extensions of Himself, a testament of who He is, surround us. Besides, He gave us the brains, senses, the conscience, and above all, the ability to choose. Why do we need a teacher, particularly in adulthood? We must go back to the drawing board – nature.

How can we have so many churches, mosques, temples, all sorts of prayer houses – all talking or teaching about the Creator (are they?) – and yet we disobey the Creator? Something must be wrong! These places of worship or prayer houses must not be in the business of teaching about God but themselves. Many signs show these institutions are not about God. First, evils and unhappiness have become the norm. There is no denial of this. Everything in nature shows that the earth was created to be a Paradise, not a place of evil or unhappiness.

Secondly, suppose these establishments are in the business of the Creator. In that case, the institutions will cater to the worshippers' physical and spiritual well-being, which will be reflected in the behaviors and attitudes of the worshippers while in or outside of the place of worship. The behavior and attitude should be observed when one is driving, in the marketplace, in the bank, in the court of law, in short, wherever one is.

It cannot be a place to show off wealth, show one is better than others, a place for gossip, or a place to plan evil against any living thing. The way we live and conduct our affairs should be our daily prayers - worshipping God. It should be a place where people are reminded of who the Creator really is, why we are created, and what our duties are here on earth. It should be a place for soul-searching to find out how one is progressing with the Creator's expectations. If one falls short, it should be a place where to renew one's commitment to do better.

It cannot be a place to spend hours or days 'praying' because one thinks the Creator needs to be reminded or coaxed or pressured or even forced to do according to one's Will. That will mean the person forgets that God is God, the Omniscience, Omnipotence, alone by Himself, needing no help. It would be easier for humans to go back to nature to see how other living things, other creations of God, deal with or handle those behaviors. Has anybody seen a goat, an elephant, a warm, a fish, a plant, a seed praying, thanking God, offering sacrifice, going on pilgrimage, etc.? Except committing evils, is there anything humans do other living things don't? What is so difficult about the Creator's expectations of us that leads us into disobedience? God does not ask of us anything beyond our abilities – we should love our neighbors (that includes all living things), as we love ourselves. People parade around, giving the impression they know and love God. Think for a moment what our behaviors will look like if we really know God and love Him. His power and glory will dazzle us to obedience, and our free will cannot interfere. His free abundance will force us to kneel whenever His name is mentioned.

There will be no time for disobedience. How could we even hurt a fly, a creature of God, the Being that created us and provided all we need? Before a simple plant is uprooted or any animal killed for any reason, we will first examine our conscience to be sure we are doing the right thing. It will be unthinkable to go into battle after receiving 'holy communion' to kill or start burning people or destroying other peoples' property after Friday prayers. How could we utter words that hurt, deceive, or cheat? How should a religious leader feel or do when any of these takes place?

We have not seen God before – we only go by what is in our imagination of Him and the perplexities of His creations that surround us. This is how God designed it. Suppose God is seen and known as a fifty-foot Giant with infinite powers, knowledge; attended to by His army of angel soldiers; can appear or disappear in any place in a moment's notice and could render judgment and dispense justice on the spot. Then a simple mention of his name would cower any person to instant obedience.

If the Creator chose this route, nothing would have stopped it. He willed instead that humans have free wills to be able to choose. For reasons best known to Him, He does not want to interfere in our choices. Yet, he is in complete control of His creations. There is hope for humans because there is a 'time' element in God's system, time for us to work out things to realize the paradise earth as He intended.

Finally, starting with the senses, the Creator provided to all animals and used most by humans: For a sense of smell, He created different types of flowers, grasses, leaves, barks of some trees, etc., with different kinds of scents to caress our nostrils and please our hearts. Children, particularly babies, can recognize their mothers through their smell. And the aromas on their loving wives' clothes can turn on many husbands.

Think of:

- Greenery, the beauty of sunrise and sunset.
- The rainbow.
- Starry nights.
- Flowers.
- Beautiful designs on fruits, leaves, animal skins.
- Smiles on the faces of a loving wife/ husband, one's children, and all the beautiful things of the earth.

God did not want His creatures, humans, to miss out on all those. Hence, He gave the sense of sight.

How refreshing and reassuring the hug from a father/mother to a grieving child, wife, husband. Even in the absence of the gift of sight, a blind person can move about, recognize, and count things using the s sense of feeling. There are many things one can enjoy only through the sense of feeling.

For humans to enjoy sweet music and melodies, God gave them the sense of hearing. How can one listen to, enjoy carry on a conversation without the sense of hearing? God knew He had to create a sense of hearing to foster communal living, cooperation, acquire or dispense knowledge. He is the All-Knowing. He wants His human creatures to enjoy themselves here on earth.

Something that looks or smells like an orange but does not taste like orange is not an orange. God knew that His animal creatures would not enjoy to the fullest His gifts of vegetables, fruits, seeds, meats, fishes, liquids, etc., without the sense of taste. And even, how can one enjoy okro soup, egusi soup, or any other soup, a cup of palm wine, or any other type of wine, fruit, etc., without the Creator's gift of taste?

How can one sit, stand, or walk or even accomplish anything without a sense of balance? Its existence is rarely recognized. It is there, though invisible. Without it, everything will be wobbly and unsafe. One cannot fix one's eyes on anything. One can't hold anything with the hands.

Think of anybody trying to read a book; the head will fall on the book. The Creator knew these difficulties and many others would occur without the sense of balance, and His creations will not be enjoyed to the fullest.

The senses take center stage of God's gifts to His human creatures. The brain could not even work – it will not have anything to interpret, no problems to solve, etc. There will be no relationships, and there will be no desires. Life will not function.

Nobody needs a crystal ball or a juju man or a diviner or laboratory experiment to tell that the intentions of the Creator in His creations were for happiness for His creatures, particularly humans here on earth. It is evident. A lot of examples have been given elsewhere in this book to show it is so. Then an overarching question: "Do you really understand who the Creator is, why you were created, the divine expectations of every human being?" These questions fall into the laps of the following categories of human beings, all humanity:

All the people in each group below (involving all the peoples of the earth) need to answer the related questions for themselves. The questions should not be taken as being judgmental. It is not meant to offend anybody. It is for self-examination to realize that our combined misdeeds make our beautiful earth an unhappy place to live. We must change our ways by reading the writings of the Creator as written in all His creations around us. Then we will know Him, obey his laws to enjoy the Paradise Earth as the Creator intended.

Imagine for a second people who see themselves as or are poor (already described before in this book). If we know who the Creator really is, our question to such persons would be, "Did you run out of the Creator's gifts – the sun and sunshine, water, air, food, natural medicine, arable land, elements that please the senses and keep us happy and make life worth living? Or did you, out of your free will, refuse to use the brain and the senses, the gifts of the Maker? Or did you refuse to "go and get" the food He has already provided . . . ."?

In this state, the 'poor' may take advantage of those who try to help them out of sympathy. The poor may feel justified to pilfer, envy, covet, etc. And above all, he may feel justified not to carry out his duty here on earth – 'take care of my earth' or rather 'live and let live.' The 'poor' must answer the same questions as anybody else.

The next group is those who have accumulated wealth and are therefore rich. They are the untouchables. With the money, you can buy almost anything! Consequently, they can do and undo. Remember, there is nothing wrong with being wealthy. After all, aren't all humans the creatures of the richest, the Creator?

However, when a person uses their wealth in competition with God's laws, problems occur.

They promote themselves to be seen by all as one of those whose will/command must be unquestionably carried out despite the use of various nefarious methods— cheating, killing, etc., involved. The person feels on top of the world while inflicting misery on many. In this state of power and glory, the primary casualty is the neglect of the Creator's law of 'take care of my earth' or 'love all my creations as you love yourself.' Those in this category are reminded of the consequences of abusing the power that wealth brings.

There is yet another group. In this group, some are 'poor,' some 'rich,' and from all races and walks of life. They believe that the earth and its free bounties belong to all humanity, current and future generations. They believe in 'one for all and all for one,' that what hurts one person hurts all and takes advantage of nobody, no matter the type of employment.

They are always in the service of all living things. The "day of judgment" for these people has already started. They are happy, enjoying the blessings of the Creator as He intended.

Another category is those in armed services, including soldiers, Police, security professionals, and law enforcement. In short, all people that carry weapons that can kill, destroy, or inflict injuries. In this employment and in possession of these weapons, some individuals feel no obligation to their natural duties to 'take care of my world or love all I created.'

They deceive themselves into believing that it is the call of duty to kill even unarmed people or persons one does/not know or had any dealings with, babies, mothers, the poor, and the weak. The faces of the children and babies grieving for losing their mothers, fathers, and property are no concern of theirs. The law of the Creator is not to destroy but to take care of all things. You have your God-given free will, but don't forget your responsibilities.

There is a category called 'the stay-at-home parent.' Many in this group work so hard, one would prefer to engage in an eight-hour job. It is not easy taking care of a single child; imagine taking care of several and, adding to that, taking care of the home itself and the individual needs of a spouse. Looking from the outside, such parents are enslaving themselves for the family. No! The parent feels happy with the duties, instinctively knowing it is a divine assignment. How can any family survive without, at the least, one of such parents? But is it better to have the two share the 'load' to lighten the weight of the responsibilities? Then they are happy doing the Creator's assignment. The "day of judgment" has already begun – the happiness they enjoy they enjoy together. There is a splinter in this group (generally women) – who feel cheated by nature for making them females and being seen or referred to as women demeaning. These are reinforced by some religious teachings that 'bearing children and toiling to care for them are punishments for disobeying God.' I would ask those teachers or peddlers what sin did all other female creatures of God commit since these also bear 'babies' and provide for them?

When a divine duty is viewed as punishment, the mind is filled with evil and revenge. It is easy to guess the consequences of such feelings—the poisonous indoctrinations fed to the children and the never-ending chaos in the family. The family may end up in a divorce, which will finally destroy all hopes and aspirations of the family members for generations. If you are unhappy executing your duties, you need to reread this book and advise yourself.

Hours of prayers, confession, making the five-a-day prayers, etc., will not do.

In many countries, primarily Western, the judgment in divorce cases is almost always in favor of the women. Generally, the woman will get the houses, cars, children, and most of the money, if any, in the bank. With a heartless partner, you can imagine the temptation to invite the legal system, knowing very well how the court will rule. Inviting the legal system is not the way to go but finding out the intention of the Creator in that, through reading and heeding his law as written in His creations.

There was a well-publicized divorce case in the US. The couple was from Nigeria, specifically Igbos. After many years in the US and receiving different degrees, the young man decided to go to his hometown for a wife. He returned with a young and beautiful one. He was satisfied and happy because no number of degrees will replace or satisfy the essential requirement that defined a man – home, wife, and children. His father was delighted and blessed a worthy son.

The traditional wedding, the white wedding, the miscellaneous expenses, and her transportation to America emptied his bank. Did he care? No! The main objective was accomplished. He became the envy of his colleagues. And two-and-a-half years into the marriage, there was a baby girl of nine months old. The man felt on top of the world. And the couple knew it was right for the woman to pursue a career in nursing.

The man's troubles started immediately after the lady received her first paycheck. Before you know it, the legal and court systems were invited into their home to settle a divorce case. I leave you with the rest of the story. Finally, the man was deported back to his home country. At this point, such women will rejoice for taking advantage of 'a stupid man.' Stupid!! She would be rolling in money, enjoying herself. Did somebody forget the Creator's law to 'take care of all I created,' or put differently, 'love your neighbor as you love yourself'?

There is yet another group: those in positions of authority. With the stroke of a pen or the click of a mouse, they can change lives or send men to battle or the gallows. In their positions, they can twist things or falsify records to transfer millions of dollars to their accounts. Armed robbers are included in this group. They have only two objectives – fame and glory. They have no time to think of the Maker of all things and His law to 'take care of all I created.' Ignorance is no excuse!

There is a group that calls itself "Men of God" or, more accurately, 'people of God.' Are there men of God? Hardly! Or point at one. If there are such men of

God, the earth would not be in the sorry state it is in, considering the number of churches, mosques, temples, etc., where teaching or preaching about 'God'; praying and worshiping go on daily. These peoples parade themselves as having powers to forgive sins; they can communicate with the Creator or serve as middlemen/women between God and humans.

Think of this for a second. If such people exist (or point at one), they will teach humans who the Creator really is. And they would obey God. Imagine what should be the reaction and the feeling of shame of such 'people of God' if a member cheats, kills, or misbehaves in any form? Is it not normal for a 'man of God' to start his preaching with the statement, "We are all sinners, and God is merciful and will forgive any sins." Is this not glorifying and encouraging sinning?

Among the members of this group are few who genuinely believe God commissioned them. Such belief would be bizarre, of course. If a person is carrying a message from the President, Prime Minister, King, or Queen of a Country to any person, there will be no doubts about whom the messenger represents – the marks of the office will be clear. If a principal or headteacher of a school sends a message to parents, there will be no doubts about who sent the message.

Then consider for a second a messenger from the Creator of all things, visible and invisible, the Alpha and Omega, the Omnipotence and the Omniscience. In short, there will be no difficulty in identifying a person sent by Him. The actions, words, or deeds of such a person will tell.

Sadly, there are four subgroups in that group: The first is the 'Conformists.' The people in this segment don't bother themselves with the trouble of using their brains and senses to examine what they hear or are told by their religious teachers. They simply followed their parents, friends, etc., blindly and satisfied.

The second subgroup has "Religion Addiction" syndrome. It does not require the consensus of wise men or laboratory experiments to understand the disastrous effects of any type of addiction – food, alcohol, drug, smoking, womanizing, religion, belief, lottery, gaming, etc. One is addicted to a behavior when the behavior takes complete control of the person. There is no need for further comments on this subgroup than to say that some people can be so addicted to abandon their family and marital responsibilities to their religious leaders or duties.

The third subgroup is the "Religion Profiteers." Some people in this group have gained power and authority over their 'flock.' They can do or acquire all their hearts' desires in the name of God. They cannot be challenged. Another section of this subgroup has amassed wealth, complete with possession of mansions, private jets, and sits on the lap of luxury. These are the teachers and leaders. It is difficult for a person in this position to tell his 'flock' who the Creator really is, His expectations, and the duties of humans here on earth. They are generally referred to as the most powerful or the richest man of God. The richest men of God, "Are you a happy man?" No!

The fourth subgroup has those who can't hear criticism on their character or conduct or their religious leaders. You can't talk about responsibility to them. They will stop associating with you if you preach to them about what God expects of us all – being your brother's keeper, the resources of this earth belong to us all. No! They can't hear bad news; they must be allowed to do as they please. These people were given their intellect, senses, conscience, etc., by the same Creator! Unfortunately, many swallowed the baits: make sure to pray five times—a day; all you need is to believe you are a son/daughter of a loving God; just say the name of Jesus or accept the blood of Jesus washes you, or all you need is the key to heaven or remember to turn the other cheek. If one sticks or buys any of these, cheating, killing, or committing any form of evil will be of no consequence to the person.

You already know, or at the least, you can imagine what these false Prophets enjoy on this earth because people are deceived into believing they possess powers they don't have.

Another group is called "businessmen/women." This group will include those who produce or sell services or information. In this group also, you will find those who received college education on how to twist minds or convince people to behave in ways; ordinarily, they would not. The main problem with this group is the quest to make more and more money and compete on who is the greatest. This inevitably leads to falsehood, cutting corners at all costs. In this type of business, there will be many promotions; there will be a lot of money to buy 'stuff' and enjoy.

The law of the Creator to love as you love yourself or take care of all He created is kept on the back burner if not forgotten entirely. Why would anybody care? Pray five times a day or, better still, say the name of Jesus. What of fasting or making a Pilgrimage? Or just say your prayers. Any of these will take care of all the wrongdoing. Some believe that when Jesus returns, they will simply take him to the back of the house and offer him some money and request he looks the other way about their evils. This is the height of absurdity.

The above analysis looked at all categories of humanity – no exceptions. Then it is time for self-examination in whatever class/category you belong to or in whichever way you conduct your business here on earth, "Are you a happy person?" The Creator provided all the resources humans need to accomplish their task of taking care of His earth or, put differently, the task of loving all He created appreciatively and happily, thereby avoiding consequences for failure. Happily! Do you see yourself as an a-once-in-a-while happy person or never?

Are you enjoying the free bounties of the Creator – water, air, sunshine, fertile soil, food, beautiful flowers, and things of that nature? Do you enjoy your family – your soulmate, children, grandchildren, parents, and grandparents? Or do you live alone? At night do you sleep with one eye open or with both closed to enjoy the Creator's time of rest? Do you lock up your house to prevent the envious from coming in to cart away what rightly belongs to you? Or do you live in a fortress to prevent the consequences of the evils you committed from reaching you?

On the streets, do you constantly look behind your shoulders or keep looking at your rear mirror? Are you edgy or restless in the restaurant or any public place or gathering, fearing someone is out to get you using any means possible? You see yourself as a poor person or not rich like your neighbor. Do you spend most of your nights angry with the world and innocent of whatever your condition is? In short, did you back yourself in a corner with greed, envy, wickedness, deceit, the desire to be on top at all costs, not caring about the Creator's law to love all His creations?

Did you forget that the Creator's law is for the good of all living things since all their welfares are tied together in an everlasting bond? Don't you wonder why you can have all money can buy yet you are ever looking for more and more, never full or satisfied? You cannot eat or use all you have accumulated or usurped; neither will you let those who can. Look at what your behavior has caused humanity – broken families, the smallest unit of our societies, polluted atmosphere, unrest, insecurity, unhappiness. Your "Last Day or your Day of Judgement" has begun? Whom are you blaming, your religious teachers, or your refusal to use your gifts of the brains, the senses, etc., to understand that all belong to the Creator but given to humans to share equitably?

You may be one of those who seek to understand who the Creator is and His expectations of us. You have been doing your best and enjoy the blessings of the Creator. You are a satisfied, happy person. These are the rewards from the Creator.

If, on the other hand, after going through the above list, you find yourself wanting, there may still be some opportunity for you to go back and undo all your misdeeds against humanity and free yourself of the self-inflicted torture. Yes! To the very persons you swindled out of their money, opportunities, starved the family, or stunted their growth.

Don't forget to include in your to-do list what you did to our climate, soil, plants, and animals if you participated or encouraged any of those behaviors unacceptable to the Creator.

What will happen to all the wealth, power, influence, glory you lied, cheated, killed, or committed all sorts of evil to acquire? On your last day on this earth, none of these things can help you. Worst of all, you will leave all behind and go as you came, alone with nothing. It does not make any sense!

The Creator's will must be done! All Glory be to Him.

ABOUT THE BOOK

Time Out: for a Second Look at Nature by Callistus Mbano engages readers with an opportunity to ask, examine, and ponder critical questions that fundamentally challenge conventional thinking and beliefs. It is natural for every normal child to be curious about these questions, but we are trained at an early age that it is forbidden to ask them. As humans, that does not stop our curiosity. We continue to seek answers from our natural environment.

Children in modern societies have fewer opportunities to play outdoors, explore nature, and ask questions of adults who ponder about life. Not examining the most fundamental questions has stifled our full human development. We go through life, confounding our reasoning with learned information built on a shaky foundation. Soon creativity and spontaneity give way to conformed behaviors.

Despite our formal education and the ways we try to impress others with our assumed superiority, many of us quiver in the face of indigenes from less developed societies when the topic discussed in Nature. When we face such people in conflicts, we become disarmed by our loss of morality.

We consider ourselves fortunate to live in the most developed technological country on earth. Chained to our devices and instruments, we are increasingly losing touch with our real selves and other people on the planet. Is it any wonder that reality TV's most popular shows are about modern people returning to primitive settings and forming tribes? Is this the only way for contemporary people to understand the basics of what it means to be human?

Amazingly in some of these reality TV shows, people continue to use the schemes and competitive methods of the modern world (in fact, the shows are competitively designed). Yet in rare situations, when modern people meet indigenes, they often find them to be communal, caring for the other even when the other is a stranger.

What if we were able to escape the 'forbidden forest' of conformity and dare to question and thoroughly examine popular beliefs and the things we were taught with our childlike inquisitiveness intact. Through reading this book, you can do this as Callistus explores the banned questions through his perspective as a youth educator and naturalist learner and teacher.

What if we were able to escape the 'forbidden forest' of conformity and dare to question and thoroughly examine popular beliefs and the things we were taught with our childlike inquisitiveness intact. Through reading this book, you can do this as Callistus explores the banned questions through his perspective as a youth educator and naturalist learner and teacher.